UNION INTERNATIONALE DES SCIENCES PRÉHISTORIQUES ET PROTOHISTORIQUES
INTERNATIONAL UNION FOR PREHISTORIC AND PROTOHISTORIC SCIENCES

PROCEEDINGS OF THE XV WORLD CONGRESS (LISBON, 4-9 SEPTEMBER 2006)
ACTES DU XV CONGRÈS MONDIAL (LISBONNE, 4-9 SEPTEMBRE 2006)

Series Editor: Luiz Oosterbeek

VOL. 20

2006
lisboa
uispp
XVcongresso

Session WS22

Theoretical and Methodological Issues in Evolutionary Archaeology

Toward an unified Darwinian paradigm

Questions théorétiques et méthodologiques en archéologie évolutive

Vers un paradigme Darwinien unifié

Edited by

Hernán Juan Muscio
Gabriel Eduardo José López

BAR International Series 1915
2009

Published in 2016 by
BAR Publishing, Oxford

BAR International Series 1915

Proceedings of the XV World Congress of the International Union for Prehistoric and Protohistoric
Sciences /Actes du XV Congrès Mondial de l'Union Internationale des Sciences Préhistoriques et
Protohistoriques
*Theoretical and Methodological Issues in Evolutionary Archaeology / Questions théorétiques et
méthodologiques en archéologie évolutive. Vol. 20, Session WS22.*

ISBN 978 1 4073 0398 7

© UISPP / IUPPS and the editors and contributors severally and the Publisher 2009

Outgoing President: Vítor Oliveira Jorge; Outgoing Secretary General: Jean Bourgeois
Congress Secretary General: Luiz Oosterbeek (Series Editor)
Incoming President: Pedro Ignacio Shmitz
Incoming Secretary General: Luiz Oosterbeek

Signed papers are the responsibility of their authors alone.
Les texts signés sont de la seule responsabilité de ses auteurs

Contacts : Secretary of U.I.S.P.P. – International Union for Prehistoric and Protohistoric Sciences
Instituto Politécnico de Tomar, Av. Dr. Cândido Madureira 13, 2300 TOMAR
Email: uispp@ipt.pt   www.uispp.ipt.pt

BAR Publishing is the trading name of British Archaeological Reports (Oxford) Ltd.
British Archaeological Reports was first incorporated in 1974 to publish the BAR
Series, International and British. In 1992 Hadrian Books Ltd became part of the BAR
group. This volume was originally published by Archaeopress in conjunction with
British Archaeological Reports (Oxford) Ltd / Hadrian Books Ltd, the Series principal
publisher, in 2009. This present volume is published by BAR Publishing, 2016.

Printed in England

BAR
PUBLISHING

BAR titles are available from:

BAR Publishing
122 Banbury Rd, Oxford, OX2 7BP, UK
EMAIL   info@barpublishing.com
PHONE   +44 (0)1865 310431
FAX   +44 (0)1865 316916
www.barpublishing.com

TABLE OF CONTENTS

Introduction.. 1
Hernán Juan Muscio, Gabriel Eduardo José López

The Application of Darwinian Cultural Evolutionary Theory to Ceramics:
The Case of "Soft Pottery" from Luwu, South Sulawesi, Indonesia................................ 3
David Bulbeck

Temporal Trends in the Morphometric Variation of the Lithic Projectile Points
during the Middle Holocene of Southern Andes (Puna Region).
A Coevolutionary approach 13
Marcelo Cardillo

Interdemic Selection and Phoenician Priesthood. Darwinian Reflections
on the Archaeoastronomy of Southern Spain.. 21
José Luis Escacena Carrasco, Daniel García Rivero

An Evolutionary Theory of Cultural Differentiation 31
Agner Fog

A Group Selection Model of Territorial War, Xenophobia and Altruism
in Humans and other Primates 35
Agner Fog

Two Faces of Darwin: On the Complementarity of Evolutionary Archaeology
and Human Behavioral Ecology 39
Kristen J. Gremillion

The Study of the archaeological record of Santa Rosa de los Pastos Grandes,
Puna of Salta, Argentina, from an inclusive evolutionary perspective........................... 49
Gabriel López

Finding Concordance in Darwinian Archaeologies: and why an Unified
Evolutionary Archaeology is both impossible and undesireable..................................... 57
Herbert D.G. Maschner, Ben Marler

The Experimental Simulation of Archaeological Patterns: A Contribution
to a Unified Science of Cultural Evolution 65
Alex Mesoudi

A Synthetic Darwinian Paradigm in Evolutionary Archaeology
is possible and convenient... 73
Hernán Juan Muscio

Niche Construction Applied: Triple-Inheritance Insights into the Pioneer
Late Glacial Colonization of Southern Scandinavia 83
Felix Riede

Acheulean Biface Refinement in the Hunsgi-Baichbal Valley, Karnataka, India................ 95
C. Shipton, K. Paddayya & M. Petraglia

Evolutionary Transitions and Co-Evolutionary Dynamics in Biology and in Culture....... 103
Mónica Tamariz

LIST OF FIGURES

Fig. 1.1. Luwu, South Sulawesi, and sites mentioned in the text 5

Fig. 1.2. Luwu soft pottery rims 7

Fig. 2.1. Different morphotypes: A 5000 B.P., B 6000 B.P., C 4000 B.P.,
   D 4000 B.P., E 6000 A.P, F 5000 B.P 15

Fig. 2.2. Upper section of the figure: Resulting NJ tree over 25 points (2 landmarks
   and 23 semilandmarks) points were previously aligned with generalized Procrustes
   superimposition procedure which was the base for the Euclidian distance matrix
   among cases. Below: expected shape to four nodes using a least square
   superposition criterion between the observed phylogeny and a very densely
   sampling outline (140 points, two landmarks and 138 semilandmarks).
   Black circles show the mean shape, while vectors and grills signal
   the mount of deformation from the mean shape 16

Fig. 2.3. A morphospace evolution model for the 6000-4000 bp time span.
   The arrows show the hypothetical direction of change thought time over
   the frequency distribution of projectile points in theoretical morphospace.
   Extreme (observed) morphologies are also depicted in the boundaries of
   the empirical distribution. Interrogation sign (?) suggest a maybe available
   -but actually- non-occupied morphospace. Relative form elongation is
   represented here like an allometric relation h/w (height/wide ratio).
   Dashed line depicts the isometric grow vector 18

Fig. 3.1. The sanctuary of Ba'al at Coria del Río (Seville, Spain) has revealed
   a clay altar in the shape of a bull skin whose longitudinal axis is directed towards
   the East to the sunrise of the summer solstice and to the West to the sunset of
   the winter solstice 24

Fig. 3.2. The Carambolo compound is also orientated towards the sun rise and set.
   Although it was originated in the 9th century BC with a more simple design,
   this solar orientation was present since its foundation and was respected in
   a later phase of enlargement 26

Fig. 5.1. Simulated evolution of group territories. Lighter colours = groups
   with higher fraction of altruists 36

Fig. 8.1. An Evolutionary Tree of Arctic House and Hearth Form 59

Fig. 9.1. The structure of a unified science of cultural evolution (right hand side),
   as mapped onto the structure of evolutionary biology (left hand side).
   Evolutionary archaeology can be seen as the cultural parallel of paleobiology.
   Adapted from Mesoudi *et al.* (2006) 66

Fig. 9.2. A schematic representation of Bettinger and Eerkens' (1999) hypothesized explanation for differences in prehistoric projectile point variation in the Great Basin. The point design of a successful hunter (A) is copied via indirectly biased cultural transmission (B), resulting in highly correlated point attributes and low variation. In guided variation (C), individuals engage in independent trial and error learning, resulting in increased diversity in point designs and low attribute correlations. In Bettinger and Eerkens' (1999) hypothesis, the stages B and C correspond to points originating from central Nevada and eastern California respectively. Mesoudi and O'Brien (submitted) simulated the learning rules in B and C (indirect bias and guided variation) and confirmed the resulting patterns of variation (low variation in B and high variation in C)............................... 67

Fig. 10.1. Evolutionary environments and artifactual variation. Circle E_a represents the selective environment at the artifact scale; circle E_i represents the selective environment at the scale of the individual organisms, and E_{ns} is a non selective surface. Horizontal line represents artifactual variation. The labels show the status of the variation as a result of its interception with each environment.................. 75

Fig. 10.2. Cladogram of the earliest ceramics of Northwestern Argentina. Each terminal taxa is a single assemblage. Mat12: Matancillas 1 and 2; Coch39: Cochinoca 39; UrcS11: Urcuro Sondeo 11; ATFIII: Alero Tomayoc Fase III; ATIV: Alero Tomayoc Fase IV; Ica1: Inca cueva Alero1; CcritCB: Cueva Cristobal Capa B; RE1: Ramadas Estructura. The information was obtained from the published bibliography (see Muscio 2004). Character states are binary, except for mean wall thickness.................. 78

Fig. 10.3. Declination throughout time of the mean wall thickness of the earliest ceramics of Argentina.. 79

Fig. 11.1. A schematic outline of the niche construction or triple-inheritance model. The total phenotype of the members of the population at t is composed of genetic and cultural components. These are connected to the succeeding descendant generation at t+1 through genetic and cultural information transmission. This part of the model is identical to dual-inheritance models of bio-social evolution. The niche construction approach, however, also recognizes a third domain on inheritance, ecological inheritance. At t, organisms from the population modify their environment so that the selection pressures exerted on them at reproduction are modified. This modified environment in then bequeathed onto the descendant generation at t+1 through ecological inheritance........................... 84

Fig. 11.2. A schematic outline of the Late Glacial in northern and central Europe. Adapted from Eriksen (1996). The three regions shown here correspond to those in Figure 11.6... 85

Fig. 11.3. A quantitative analysis of curation in scraping tools of the Late Glacial forager groups, following the method presented by Shott (1993), finds no evidence for higher curation rates in the Hamburgian. The total sample of 1577 specimens measured by the author is divided by cultural group/site context (as indicated by the excavator or curator). In support of the view that Hamburgian tools may have been maladaptive in the Southern Scandinavian context, Kuhn (1994) suggests that the ideal size of scraping tools should be only 1.5 times their length. Remarkable, the Federmesser groups are distinguished by scraping tools (so-called thumb-nail scrapers) that fit this prediction reasonably well 87

Fig. 11.4. A maximum likelihood tree of 16 Late Glacial projectile point taxa juxtaposed to key stratigraphic sequences from Ahrenshöft, Rissen, Bettenroder Berg and Bad Breisig and the $\delta^{18}O$ temperature proxy data of the GISP2 ice core. LST = Laacher See-tephra 87

Fig. 11.5. Eriksen's (1996) map of the European Late Glacial, which corresponds to Figure 11.2. The Laacher See-tephra cuts Southern Scandinavia off from the core settlement areas. Interestingly, the Thuringian Basin (2) appears to have been depopulated prior to the Laacher See-eruption, leaving this key area as a Palaeolithic no-man's land after groups moved away from regions affected by ash fall-out. Cultural historical schemes suggest that even in south-central Europe, the demographic fluctuations instigated by the Laacher See-eruption may have led to (largely stylistic) cultural changes 88

Fig. 11.6. The inconsistency in the classification of Federmesser and Bromme points in Table 1 is resolved when the measurement distributions of the pivotal variable, maximal width, is plotted as a simple histogram. The Federmesser sample is bimodal with the subsidiary mode representing large tanged points found in Federmesser contexts. These indicate the use of a secondary weapon delivery system (the dart and spear-thrower) in addition to the bow-and-arrow on the Northern European Plain prior to the Laacher See-eruption. After the eruption, the bow-and-arrow disappears from the repertoire and is replaced by the exclusive use of darts. Those Bromme specimens identified as arrows either denote a late trend towards arrow tips or a misidentification of these points inherent in the method used 89

Fig. 12.1. Thickness to breadth ratio by site 98

Fig. 12.2. Mean biface weight in grams by site 98

Fig. 12.3. Mean flake scar area by site in mm^2, for the last four flake scars removed from the piece.. 99

Fig. 12.4. Flake scar width to length ratio by site for the last four flakes removed from the bifaces.. 99

Fig. 13.1. The elements of a Darwinian selection system................................. 104

Fig. 13.2. Elements and mechanisms of selection of public forms.................... 106

Fig. 13.3. Elements and mechanisms of selection of cultural competences....................... 107

LIST OF TABLES

Tab. 1.1. Differences between sites in their proportions of higher-fired "Pink" and "Orange", and the ratio (by weight) of higher-fired shards to soft pottery 8

Tab. 1.2. Recorded weights (in grams) of soft and non-soft pottery at Luwu soft pottery manufacturing sites. The upper half at Utti Batue includes spits 5 – 10, and the lower half includes spits 13 and 14 (time did not permit analysis of the other excavated spits) 9

Tab. 4.1. Regal and kalyptic characteristics in various spheres of life 32

Tab. 7.1. Radiocarbonic dates of Alero Cuevas site 50

Tab. 11.1. The classification of the measured Late Palaeolithic projectile points following Shott's (1997) discriminant function analysis (one-trait variant using width) and using measurements taken by the author on complete or nearly complete specimens. Note that arrow shafts are known from Ahrensburgian contexts. The methods ability to correctly classify these projectiles as arrow heads lends confidence that the other assessments are also largely correct (the method is known, however, to underscore dart points) 89

INTRODUCTION

Hernán Juan MUSCIO and Gabriel Eduardo José LÓPEZ

Grupo de Investigación Cultura, Comportamiento y Evolución Humana (GICCEH), Sección Arqueología, Universidad de Buenos Aires. CONICET, 25 de Mayo 217 3° Piso, Buenos Aires, (1002) ARG, w22_archaeoevolu@yahoo.com.ar

This volume presents the contributions to the *Workshop 22: Theoretical and methodological issues in evolutionary archaeology: toward a unified darwinian paradigm,* xv[th] *Congress of International Union for Prehistoric and Protohistoric Sciences.* The workshop was proposed to discuss some important theoretical and methodological issues in Darwinian evolutionary archaeology.

In the beginning of this century, Darwinian evolutionism has become the only theoretical coherent framework in scientific archaeology. Still, today, several lines of Darwinian archaeological research coexist without a full integration. The absence of integration between theoretical richness slows down the emergence of a solid paradigm regarding the causes and consequences of the evolutionary change, as it is perceived in an archaeological record. This is the case of the application of the Human Behavioural Ecology, Cultural Transmission Theory, Human Socio-biology, Evolutionary Psychology and the Evolutionary archaeological. Recent important books as *Systematic and Archaeology; Genes, Memes and Human History: Darwinian Archaeology and Cultural Evolution; Cladistics and Archaeology, Archaeology as a Process,* and our own investigation in Argentina convinced us that the integration is possible and theoretically convenient. Our suggestion is that every Darwin inspired theoretical framework can contribute to the evolutionary archaeology so long the evolutionary processes and the material results proposed are recognizable in the temporal and spatial scales, proper of archaeology. As an example we proposed that optimizing natural selection, was the mechanism that favoured the economic intensification of the camelids in the Andes of Argentina, during the Holocene. Also, Muscio (2004) for the same region argued that optimizing natural selection favoured thin wall vessels during the late Holocene. In our discussion optimal foraging is used as a framework to build hypotheses about the action of selection over large temporal scale.

With these ideas in mind, the meeting was thought in order to generate a rich discussion regarding the theoretical and methodological issues necessary to the integration of the several evolutionary research programs in archaeology (see for example Gremillion; López; Maschner and Marler, Mesoudi in this volume).

To accomplish this goal prominent scholars working in a wide range of time periods, geographic areas and theoretical issues contributed at the meeting. At the core of the debate were questions such as the units of selection, the role of the cultural transmission, the construction of cultural lineages, the linkages between adaptive ecological behaviour and the broad time scale processes from which emerge archaeological patterns, between others. This diversity of perspectives is evident in the present volume.

David Bulbeck's paper allows explaining in a darwinian perspective the introduction and expansion of "soft pottery" in a early state in Luwu, Indonesia. It is a interesting contribution for understanding the change in the ceramic technology and the social change in general inside of a darwinian perspective.

Marcelo Cardillo presents an important theoretical and methodological contribution for explaining the change in lithic artifacts of the Salta's Puna, Argentina. In this sense, he introduces the geometric morphometric, for discuss its use in phylogenetic analysis since a coevolutionary perspective.

José Luis Escacena Carrasco and Daniel García Rivero shows as a darwinian perspective and concepts as interdemic selection can be applied to the study of complex societies as the Phoenicians. In this case, they study the religion and the astronomical knowledge owned by the phoenician priesthood as a form of adaptative mechanism that promoted the demographic growth and geographic expansion.

Agner Fog presents two papers that explain different forms of the cultural evolution. In his first paper, Fog introduces the r/k theory which allow predicting cultural differences and archaelogical expectatives. The second paper is a interesting theoretical and methodological approach for the study of the emergence of behaviors related to mechanisms of group selection. This is analysed with computer simulations.

Kristen Gremillion shows as two evolutionary perspectives, Human Behavioral Ecology and Evolutionary Archaeology (Narrow Sense), can be two forms complementaries of explaining inside of a darwinian approach. She shows that "the ecological-functional approach of Human Behavioral Ecology and the more historical emphasis of Eolutionary Archaeology Narrow Sense are for the most part complementary – representing the two faces of Darwin". Therefore is a very important contribution for achieving a integration in evolutionary archaeology (in wide sense).

Following this vision, Gabriel López proposes an inclusive evolutionary perspective for the study of the archaeological record of Pastos Grandes, Puna of Salta, Argentina. In this case he uses two complementary perspectives: Evolutionary Ecology and Cultural Transmision Theory.

Herbert Maschner and Ben Marler discuss the difference between inclusivity and unification. They think that the better way in evolutionary archaeology is the inclusivity. They consider to the inclusivity like "a multiplicity of approaches which share a common goal" and they argue that "diversity is good for science and good for knowledge". Also they exemplify the proposal with evidence of the Alaska Peninsula in the western Gulf of Alaska.

Alex Mesoudi analyzes the potential of the simulation for predicting patterns of cultural transmission in the archaeological record. He proposes that "experimental simulations can be used in conjunction with archaeological and ethnographic studies"... "to provide a more complete explanation of past cultural change than any one of these methods alone", regarding archaeology within a larger science of cultural evolution and a unified Darwinian evolutionary approach to human culture.

Hernán Muscio presents a synthetic Darwinian framework, with the potential to link the short term microevolutionary mechanisms to the macroevolutionary patterns documented in the archaeological record. After focusing on some of the elements of a multilevel Darwinian framework for evolutionary archaeology, he discusses the evolution of the earliest ceramics of Northwestern Argentina.

Ceri Shipton, K. Paddayya and M. Petraglia show the importance of the study of the variability within a evolutionary perspective in the Acheulean technology. They present two hypothesis for explaining the variation recorded: "differing levels of biface rejuvenation between assemblages or different socio-cognitive capabilities between the hominins responsible for manufacturing the different assemblages"

Felix Riede presents a successful use of the niche construction model in archaeology, particularly in Late Glacial re-colonization of Southern Scandinavia. This model allows understanding the coevolution between genes, behaviors and environments, considering principally the ecological inheritance inside of this inclusive evolutionary perspective.

For last, Mónica Tamariz presents a coevolutionary perspective for explaining the cultural evolution. She says that "culture comprises two kinds of information: neurally-encoded private meanings and information about the structure of public cultural forms". Also she proposes that "the evolutionary dynamics of the systems of cultural forms and competences are analogous in some fundamental ways to molecular and organismal evolution in biology". Therefore is a very interesting proposal for to study the cultural change.

THE APPLICATION OF DARWINIAN CULTURAL EVOLUTIONARY THEORY TO CERAMICS: THE CASE OF "SOFT POTTERY" FROM LUWU, SOUTH SULAWESI, INDONESIA

David BULBECK

School of Archaeology and Anthropology, The Australian National University

Abstract: *"Soft pottery" constitutes a distinctive class of earthenware at major habitation sites associated with the early Bugis state of Luwu. It has many unusual features such as low firing temperature, irregular surfaces, and textile impressions on the interior surface. The evidence from the shards, that this was makeshift pottery of poor quality, has been difficult to reconcile with the evolution of complex political organization in Luwu concomitant with the production of soft pottery. Application of a Darwinian perspective, however, allows the temporary popularity of soft pottery to be explained in terms of meeting a sudden hike in local society's demand for domestic ceramics.*
Keywords: *Darwinian cultural evolution, market forces, ceramics, Luwu, Bugis*

Resumé: *La "poterie molle" est une classe distinctife de las céramiques locales laquelle se rencuentre aux sites majeures lesquels appartenent au royaume ancien et Bugines de Luwu. Cette poterie-ci se fit cuire à feu doux, et montre beaucoup des traits peu commun comme les surfaces irrégulières et las impressions des textiles pour la surface intérieure. C'était difficil de réconciler cet évidence de cette poterie improvisée, et de maigre qualité, avec l'évolution de la politique organisation complexe dans Luwu pendant la période de la fabrication de la poterie molle. Néanmoins, l'application d'une perspective darwinniène explique la popularité temporaire de la poterie molle sous l'angle de l'augmentation soudaine de la demande locale pour las céramiques domestiques.*
Mots-clés: *l'évolution culturele darwinniène, forces du marché, las céramiques, Luwu, les Bugis*

INTRODUCTION

In his major book *Ceramic Theory and Cultural Process*, Arnold (1985) comprehensively reviewed the literature from experimental archaeology and ethnography on earthenware technology, production and use. Arnold recognized the role of the ecological prerequisites for earthenware production – suitable clay, combustible material, and spells of fine weather for firing vessels – which cultural ecology has emphasized. However, after developing a systems model built around society's demand for ceramics, and the role of land shortage in promoting specialist craft, Arnold focused on social complexity as the critical factor to explain the scale and scope of a traditional ceramic industry. His concluding advice to archaeologists combined theoretical admonitions, such as his rejection of culture history's tendency to detach ceramic attributes from their social context, with practical suggestions, such as his recommendations to focus on vessel shapes, the temporal persistence of vessel-forming techniques, and fabric analysis.

From the point of view of Darwinian archaeology, Arnold's (1985) systems approach, with its terminology of regulatory feedback and deviation amplifying mechanisms, is more a metaphor than a well-founded explanatory model. Darwinian evolutionary theory is based on the changes over time in the relative frequencies of heritable traits within a population, and the relative successes of populations, through natural selection. This "survival of the fittest" formula says nothing in particular on the generation of novel heritable traits, but a pure Darwinist perspective holds that novelties arise randomly with respect to the evolutionary trajectories that result from natural selection (Rindos, 1986). In Darwinian

terms, Arnold's "regulatory feedback" would apply to heritable traits with an optimized expression, causing variation away from this optimal expression to be selected against, whereas "deviation amplification" would be expected in cases of traits whose expression has become sub-optimal due to change.

To be sure, the application of Darwinian theory is very broad and can focus on the lineages (and their relationships) along which traits are transmitted (e.g., Cameron and Groves, 2004), the development of complex structures through evolutionary tinkering (e.g., Lieberman, 2006: Chapter 6), cumulative change independent of fitness as the result of drift, the creation of a daughter population very different from its parent owing to the founder effect (Wright, 1968-1978), or niche construction (Odling-Smee *et al.*, 2003). Cultural evolutionary theory not only shares these perspectives with biological evolutionary theory, but also needs to fully consider the potential for heritable traits to transmit horizontally across lineages (Boyd and Richerson, 1985). Culture historians working in a phylogenetic paradigm, culture-contact theorists who stress reticulation, cultural ecologists who focus on adaptation, cultural materialists, multilinear cultural evolutionists and of course cultural selectionists can all take heart from Darwinian theory, and continue their unresolved, theoretical disputes with each other.

What is the point of a Darwinian perspective if it doesn't change archaeologists' theoretical penchants? My response would be that a Darwinian perspective counsels archaeologists to be both more disciplined and more inclusive in the narrative explanations (i.e., cause and effect scenarios; see Hausman, 1998) which they develop for their case studies.

The discipline comes from restricting explanations for change to transmittable traits. Ideas and "cultural baggage" no doubt exist, in some sense, but their transmission is problematic. Practices on the other hand are eminently transmittable, being conveyable through the spoken word and especially through demonstration, most effectively when tied to the contexts where they apply. Bourdieu's concept of habitus neatly encapsulates the appropriate bounds for culture history, both in terms of the patterned practices that people become accustomed to, and their collective scope for sharing (transmitting) these practices (Whittle, 2003). Similarly, artefacts do not constitute lineages in their own right; the real lineages are the practices involved in making artefacts, even when these practices are transmitted horizontally through instruction or emulation. However, if a Darwinian perspective eschews idealist theoretical positions for their lack of an anchor in the material world, it would be equally critical of systems theory and adaptationist agendas (which I call "abstract materialism"). How a society holds together and how particular practices prosper are the outcomes of natural selection, not the pre-ordained goals of some nameless social engineer.

The inclusiveness of Darwinism comes from the recognition that social continuity and social change are dynamically interwoven. Ethnography and archaeology both strongly indicate that some populations grow while others decline, communities invariably interact, practices are passed on with varying degrees of authenticity and persistence, innovation (within social and technological limits) is chronic, and people adapt to their environment but imperfectly so. Any archaeological explanation that fails to attend to these factors tells us more about the archaeologist's personal agenda than the case study the archaeologist is supposedly addressing. I suggest that the strong promulgation of either an idealist or an abstract materialist theoretical model presupposes "purification" of the available information to force fit the case study into a partial and partisan world view. To summarize, if the archaeologist focuses on cultural traits that can be both practised and transmitted, then the explanation for change will naturally move to a broad-based, holistic account.

Systems theorists may well contend that their goal is precisely a broad-based, holistic account, and they could point to Arnold's (1985) study in that context. My objection is that useful archaeological frameworks are very rarely systems, in the sense understood by engineers or information technologists. D.H. Thomas's (1972) computer simulation of the Western Shoshone economic cycle may well be a true system, but the textbook examples of "systems" in archaeology, featuring diagrams with captioned boxes linked together by arrows (e.g., Renfrew and Bahn, 2000: 471-485), are not. Exemplary archaeological frameworks, yes, but not systems. For Darwinism to contribute to the future development of archaeological theory, it should be able to start with the cause-and-effect frameworks of the sort depicted by Renfrew and Bahn (2000: 471-485) and sharpen their

explanatory value. An important concept here is market mechanisms, whose importance for traditional ceramics is specifically recognized by Arnold (1985), and which should act as a vehicle for natural selection in any industry with distinct producers and consumers.

The scope for market mechanisms to fine-tune production in a viable industry, quash an industry which is non-viable, and link product diversity to consumer power, should be obvious. Pottery production, for instance, involves considerable costs in labour, materials and storage, while potters who meet their subsistence needs by trading their pots will be only too aware of product lines that fail to attract customers. An economy hardly needs to conform to the canons of classic microeconomic theory for wasted effort to bite hard into poor production schedules, or for unsatisfied demand to stimulate new entrants. Attention to market mechanisms is arguably a very underdeveloped component of current cultural Darwinian theory. In one of the few papers addressing the topic, Boyd and Richerson (2005) explain how Darwinian theory can provide a deep theoretical basis to microeconomic theory, which is fair enough, but there should be a complementary recognition of market mechanisms as an efficient vehicle of natural selection.

This paper addresses a peculiar feature of the archaeological record of Luwu, in South Sulawesi (Figure 1.1), during its "pre-Islamic period" between the thirteenth and early seventeenth centuries CE. This feature, "soft pottery", is temporally and spatially associated with the expansion of Bugis speakers along the northern rim of the Gulf of Bone, and their establishment of a state-level organization which, during its heyday, was the most powerful polity in South Sulawesi (Bulbeck and Caldwell, 2000; Bulbeck et al., 2006). From a naïve culture history viewpoint, the association of soft pottery with a dominant immigrant population might suggest its introduction by the Bugis. From a similarly naïve multilinear cultural evolutionary perspective, soft pottery's association with the formation of a complex society would hint at craft specialization or an advanced technological capacity. The available information, however, strongly suggests that the soft pottery had local origins, and that it was second-rate, makeshift pottery. These counter-intuitive findings can be explained, as suggested here, in terms of the stimulated production of an inferior product to meet demand that was not otherwise being satisfied through ceramic production.

PRE-ISLAMIC LUWU

Based on research by the "Origin of Complex Society in South Sulawesi" (OXIS) project, Malangke (Figure 1.1) can be identified as the pre-Islamic capital of the Bugis kingdom of Luwu. Luwu was the first South Sulawesi kingdom to officially convert to Islam, in 1605, and the tombs of Luwu's first two sultans are located in Malangke. During the preceding centuries the Luwu

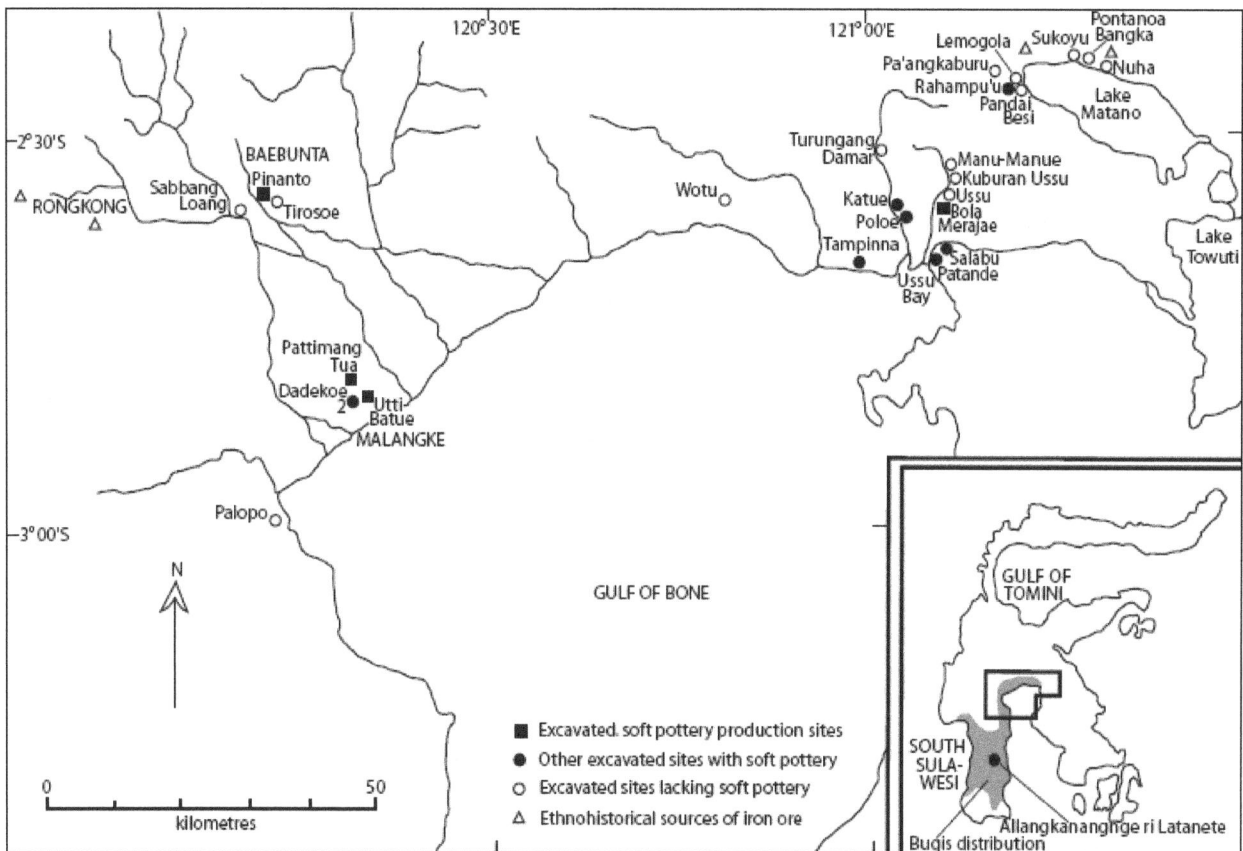

Fig. 1.1. Luwu, South Sulawesi, and sites mentioned in the text

Bugis had cremated the deceased and buried the ashes inside large jars along with a wealth of metallic and ceramic goods. Malangke's pre-Islamic cemeteries have all been thoroughly looted, but locals remember the location and extent of the looted areas, while the age of the burials can be gauged from the imported ceramics (Chinese, Thai and Vietnamese) still held in villagers' homes or represented by surface sherdage. Based on the archaeological survey of looted burial grounds, the population of Malangke is estimated to have risen from approximately 2.700 to 14.500 persons between the fourteenth and sixteenth centuries. A maximum population size over 10.000 is supported by the recorded areas of 5.3 and 4 hectares for Malangke's two main settlements, Pattimang Tua and Utti Batue. In the early seventeenth century the Luwu royalty relocated the capital to Palopo, and Malangke then lay abandoned until its re-occupation in the last few decades by cash croppers growing mandarins and cacao (Bulbeck, 2000; Bulbeck and Caldwell, 2000).

Malangke lies in the floodplain created by the Rongkong and Baebunta rivers which originate in rugged highland country (Figure 1.1). These rivers meet the coastal plain in a region traditionally inhabited by the Lemolang, whose language is very different from Bugis (Grimes and Grimes, 1987). In contrast to Malangke, which appears to have been vacant before the fourteenth century, Baebunta (as the Lemolang polity was called) has witnessed two

millennia of occupation. Habitation debris including iron artefacts at the site of Sabbang Loang are firmly dated to the early centuries CE. The source of iron is not known but may have been the Rongkong highlands where "weapons grade" iron ore was quarried and smelted in historical times. During the pre-Islamic period, the major Lemolang settlement was Pinanto, which extended 0.6 hectares along a ridge overlooking a looted area approximately one hectare in area. The close relationship between Baebunta and Malangke is reflected, *inter alia*, by Baebunta's adoption of Islam in the same year that Luwu converted (Bulbeck, 2000; Bulbeck and Caldwell, 2000).

OXIS also focussed on Ussu Bay, at the northeast of the Gulf of Bone, where several rivers converge on tidal mangrove forest. Linguistically the mainstream language of this area is Padoe, whose speakers extend eastward to the Matano and Towuti lakes (Grimes and Grimes, 1987), but the *To Ussu'* (Ussu people) constitute a distinctive Bugis enclave. Numerous sites along the Ussu River, and the Cerekang River immediately to the west, have mythical associations with the origins of the Bugis and are barred from entry. These forbidden sites probably coincide with pre-Islamic sites because extensive exploration along the Ussu and Cerekang rivers found pre-Islamic sites to be elusive, whereas they were readily located in the near environs. However, excavations adjacent to two sacred sites proved to be unexpectedly

5

rewarding. The two test pits at Bola Merajae yielded little else than pottery, but the radiocarbon dates indicate two periods of habitation, corresponding to the first millennium CE and the fourteenth to seventeenth centuries respectively. Katue has been interpreted as a riverside settlement inhabited during the first millennium CE, but my subsequent analysis of the potshards in two test pits abutting the main site indicates light occupation during the pre-Islamic period (Bulbeck and Caldwell, 2000; Bulbeck, in prep.).

Local history (e.g., Pelras, 1996) intimately links the *To Ussu'* to the trade of iron wares from Lake Matano, the source of the *pamor luwu'* prized for kris (dagger) production in medieval Java. The excavations performed by OXIS investigated five iron smelting sites identified by their concentrated debris of iron ore waste, iron slag, charcoal and baked sediment. Iron smelting had commenced by 900-1000 years ago at Sukoyu and Nuha, on the northern shore of Lake Matano, and continued till the eighteenth century at Nuha. The major smelting deposit was found at Matano, on the lake's western corner, where it is dated to between the fifteenth and seventeenth centuries at two sites (Rahampu'u and Pandai Besi); the focus of iron smelting then moved a short distance north to Lemogola. Bulbeck and Caldwell (2000) suggest that iron from Matano's northern shore was exported northwards via the Gulf of Tomini in the early second millennium CE, before being exported eastwards through Matano (and Ussu) by the fourteenth century, after which Matano became the major smelting centre.

The plans by OXIS to excavate at Rongkong were unfortunately scuttled by the destruction of the road up to Rongkong, but it would be reasonable to assume that its iron industry followed a similar pattern of development to that at Lake Matano. Thus, the combined historical and archaeological evidence strongly implies that the iron trade underpinned Malangke's (Luwu's) burgeoning prosperity between the fourteenth and early seventeenth centuries. Malangke flourished as the entrepôt for iron transported downriver from Rongkong and coastally from Ussu Bay. Indeed, the decision by the Luwu royalty to relocate the capital to Palopo in the early seventeenth century would appear to reflect the economic decline in the importance of Luwu iron, owing to factors such as the late sixteenth century introduction of firearms to South Sulawesi, and the growth of organized iron-working operations in the major population centres south of Luwu (Bulbeck and Caldwell, 2000).

SOFT POTTERY IN LUWU

Luwu's soft pottery is identified by the low-fired status of the vessels, the rounded edges of the shards, and a fabric that looks silty to sandy in texture and seems low in inclusions (around 0 – 3% of the fabric). The shape of the pores resembles stalks and irregular granules, suggesting the inclusion of soft vegetable matter that had burnt out

during firing (see Figure 1.2(d) and 1.2(f)). Except at Pinanto (see below), gleaming specks which look micaceous are the most common mineral inclusion, followed by rounded, white and black grains. Both the interior and exterior surfaces tend to be irregular, with dimples, creases, gashes and asymmetric bosses. A common feature is a thin greasy covering that occurs irregularly on the exterior and/or interior surface (see Figure 1.2(a)). Lampert's (2003: 213) chemical analysis of this slip on the Bola Merajae shards suggests the trace inclusion of dammar gum from the *Agathis* pine which occurs in the Luwu highlands.

Soft pottery has Munsell colours which are quite distinct from the standard browns and reddish browns of most Luwu pottery. Based on Munsell colour, soft pottery can be subdivided into "soft white", with white to light grey coloration, "soft pink", with pink, pinkish grey and light reddish brown coloration, and "soft orange", where the Munsell colours are typically reddish yellow and yellowish red. The more comminuted, rounded, malformed or lower fired shards of soft pottery can be difficult to distinguish from sediment clods, particularly the lumps of baked sediment frequently excavated in Luwu sites. At the other extreme, soft pottery grades into higher fired pottery of similar colour (except for the lack of a white variant), fabric and shape, such as the two examples illustrated in Figure 1.2(e) and 1.2(k). The higher firing correlates with a lesser propensity for the shard walls to be rounded, and the occasional presence of a reduced core contrasting with the vessel's oxidized walls.

The macroscopic differences between the shards at the four sites with the highest concentration of soft pottery suggest local manufacture.[1] Utti Batue is the only site to yield soft white, and soft pink is approximately ten times more common than soft orange. Textile impressions, whether pointillist (Figure 1.2(e)) or cross-hatched in appearance (Figure 1.2(c), (d) and (f)), commonly occur on the interior surface of the Utti Batue examples.[2] The Bola Merajae soft pottery resembles its Utti Batue counterparts in the frequent occurrence of internal textile impressions (Figure 1.2(a)), but the colours are very different, with soft orange about seven times more common than soft pink, and no soft white. The Pattimang Tua soft pottery is also distinct from the Utti Batue soft pottery, despite these sites' proximity (Figure 1.1). Internal textile impressions occur very rarely (see Figure 1.2(b) for one of the few examples), and the pottery is evenly divided between soft pink and soft orange. Finally, at Pinanto, no textile impressions were observed, soft orange was approximately four times more common than soft pink, porous pseudomorphs from burnt-out vegetable matter were not observed, and the most common inclusions were reddish granules dissolving into the general matrix. The higher fired counterparts of the soft

---

[1] Chemical analysis of the fabric is yet to be performed.

[2] See Bulbeck *et al.* (2006) for details. Note that the classification of the Utti Batue earthenware has been updated since that paper was written.

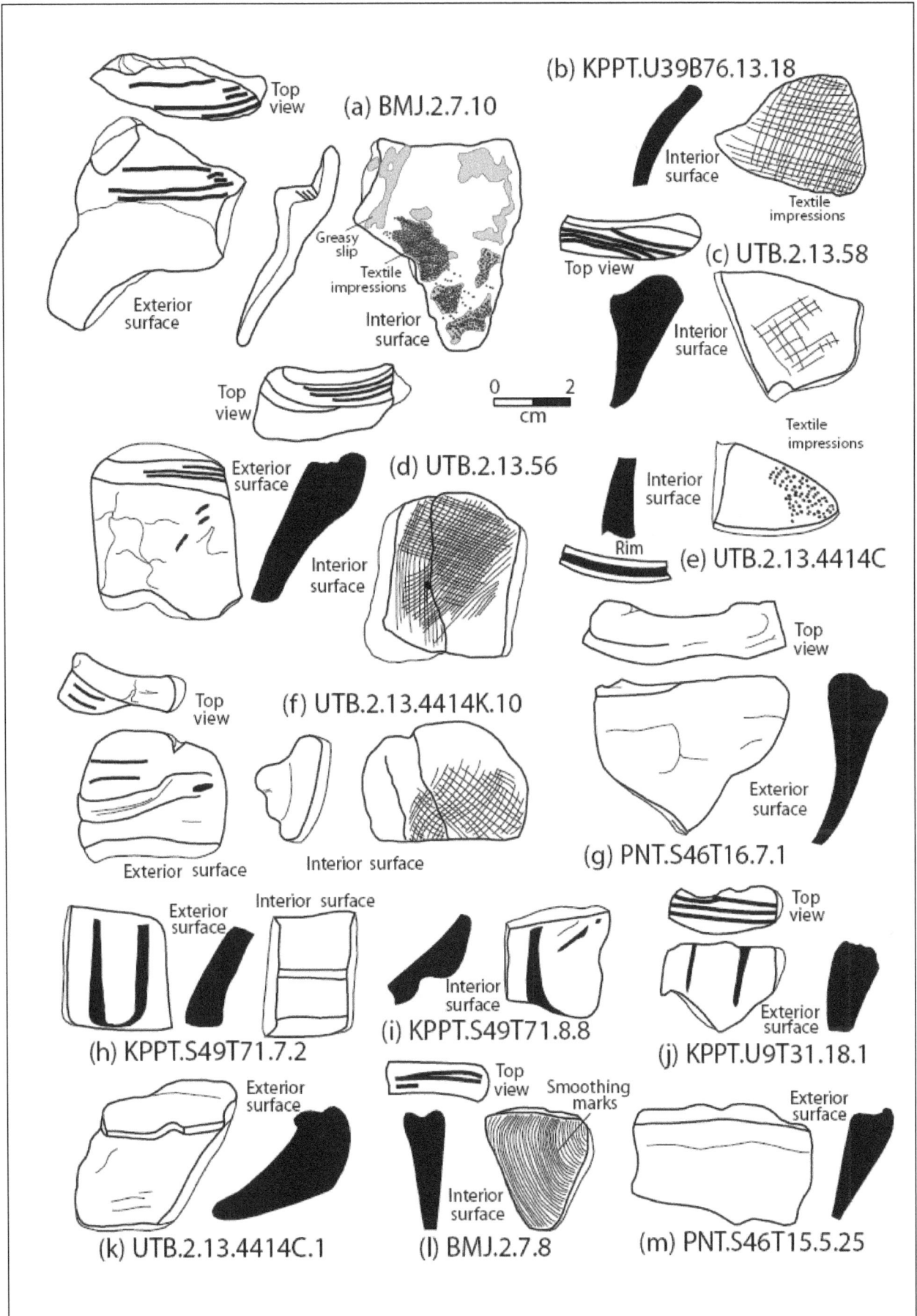

Fig. 1.2. Luwu soft pottery rims

Table 1.1. Differences between sites in their proportions of higher-fired "Pink" and "Orange", and the ratio (by weight) of higher-fired shards to soft pottery.

Site	Dominant Munsell colour class (none is whitish)	Ratio compared to soft pottery
Utti Batue	"Pink" twice as common as "Orange"	~ 1:4
Pattimang Tua	"Pink" slightly more common than "Orange"	~ 1:2
Pinanto	Almost entirely "Orange"	~ 1:10
Bola Merajae	Almost entirely "Orange"	~ 1:12

pottery also appear to differ between manufacturing sites (Table 1.1).

Both the soft pottery and its higher-fired counterparts can be dated to between the fourteenth and seventeenth centuries (Bulbeck and Caldwell, 2000; Bulbeck, in prep.). This is the age range of the great majority of the imported ceramics from Utti Batue, Pattimang Tua and Pinanto, supported by a radiocarbon date of 390±90 BP (AD 1400 – 1670 cal. at two sigma) from Pinanto. This is also the inferred age of the Bola Merajae examples, which occur above a date of 1260±60 BP (AD 980 – 1260 cal.), and in association with dates of 580±70 (AD 1284 – 1438 cal.) and 310±40 BP (AD 1480 – 1670 cal.). Similarly, at Luwu sites where soft pottery occurs at low amounts (1 – 3% of sherdage by weight), the assemblage is usually dated to between the fourteenth and seventeenth centuries by the associated imported ceramics, radiocarbon dates, or both. This is the case at Dadekoe 2, Tampinna, Patande, Salabu, and Rahampu'u (Figure 1.1). Katue and Poloe are the sole exceptions: the Katue test pits with soft orange (adjacent to the main site) are undated; and the imported ceramics at Poloe date to between the seventeenth and nineteenth centuries, though in this case they appear to have been deposited as complete vessels in an older habitation deposit. Overall, soft pottery's regular recurrence in habitation deposits dated to between the fourteenth and seventeenth centuries, and its effective absence from earlier or later habitation deposits, recommend soft pottery itself as a chronological marker of Luwu's pre-Islamic period.

The possibility of chronological change in the frequency of soft pottery, during the pre-Islamic period, was raised by my observation in two of the Pinanto test pits (S46T16 and U46B7). Here, the lowest spits included more or less equal amounts of soft and non-soft pottery, whereas non-soft pottery was dominant in every other Pinanto context. To test the hypothesis of a decrease in the frequency of soft pottery over time, I divided the excavated test pits with soft pottery into upper and lower halves. The higher-fired counterparts of soft pottery were excluded from analysis because, for some of the Pattimang Tua and Pinanto test pits, time permitted only a rough and ready classification of the shards into soft and non-soft. Where an odd number of spits in a test pit were included in the analysis, the middle spit was assigned to the upper or lower half depending on which assignment most evenly

distributed the pottery between the two halves. Body weights rather than shard counts were used for pottery quantification because the soft pottery shards tend to be smaller and lighter than the other shards. The shard weights (soft and non-soft pottery) for the upper and lower spits of each test pit at a site were then aggregated to represent the "upper" and "lower" shard weights for the site as a whole.

As indicated in Table 1.2, soft pottery constitutes a larger proportion of all pottery in the lower half of all four analysed sites. This holds true whether the ceramic assemblage is dominated by soft pottery (as at Bola Merajae), or whether soft pottery constitutes less than ten percent of the assemblage (as at Pattimang Tua). Overall, soft pottery appears to have played a decreasing role in Luwu earthenware assemblages during the pre-Islamic period, a trend that continued with its disappearance following Luwu's abandonment of Malangke as its capital.

What role did soft pottery play in the Luwu earthenware assemblages? The sharp concentration of soft pottery at four sites, and the differences between these sites, suggest that, in accord with the vessels' fragility, consumption was largely restricted to manufacturing location.[3] Moreover, analysis of the vessels' form and use context (cf. Arnold, 1985: 234-237) suggests a primary use in a domestic context, probably for serving food which could then be distributed amongst the diners.

As regards form, Bulbeck et al. (2006) identified inverted jars (Figure 1.2(a), (b), (c), (d), (f), (g) (k)), lids (Figure 1.2(e), (h), (i)), and possible boxes (Figure 1.2(l)) as the dominant vessels. Indeed, there may have been only a single vessel form – inverted jars, often with lids. The curvature of the rim tends to be irregular (see Figure 1.2(a), (c), (d), (f), (g), (j)), and this irregularity could make a short segment from a curved rim appear straight. This is particularly true because the aperture diameter, whilst never measurable, probably tended to be large (over 20 cm) given the typically modest degree of rim curvature. A wide rim aperture would also have let the potters impart textile impressions on the interior surface;

---

[3] Bulbeck et al. (2006) suggested the use of soft pottery in transporting goods across the landscape of Luwu, but that paper was written prior to thorough study of the relevant assemblages.

Table 1.2. Recorded weights (in grams) of soft and non-soft pottery at Luwu soft pottery manufacturing sites. The upper half at Utti Batue includes spits 5 – 10, and the lower half includes spits 13 and 14 (time did not permit analysis of the other excavated spits).

	Bola Merajae	Pinanto	Utti Batue	Pattimang Tua	All four sites
Upper half – soft pottery	32.5 g (67.3%)	1,004 g (12.7%)	310.1 g (6.6 %)	296.3 g (2.3%)	1,642.9 g (6.5%)
Upper half – non-soft pottery	15.8 g (32.7%)	6,888 g (87.3%)	4,354,7 g (93.4%)	12,357.7 g (97.7%)	23,616.2 g (93.5%)
Lower half – soft pottery	501.8 g (84.2%)	1,124 g (14.7%)	608.4 g (12.9%)	910 g (10.5%)	3,144.2 g (15.2%)
Lower half – non-soft pottery	94.4 g (15.8%)	6,532.8 g (85.3%)	4,123.6 g (87.1%)	6,832 g (89.5%)	17,582.8 g (84.8%)

for instance, by wrapping cloth around the anvils to cushion their effect on the soft pottery during paddle and anvil finishing. The rims are typically thickened compared to the shoulder, and often have grooves or furrows along their top surface (Figure 1.2(b), (j), (l)), or exteriorly lateral flanges which carry irregular sets of grooves (Figure 1.2(a) and (f)). These rim elaborations would have helped lock the lid onto the rim. At the Utti Batue test pit, in spit 8, two flanged and furrowed rim and lid shards, locking into each other, were recorded. Traces of decoration are sparse, with only two identified examples (Figure 1.2(h) and (k)), and even these cases could be manufacturing defects that mimic decoration. Finally, given the rim thickening and weak constitution of the fabric, it is likely that the jars were squat, because if the jars were tall, the thickened rims would have tended to collapse under their own weight. In summary, the soft pottery vessels were probably wide-mouthed jars with weakly inverted rims, and oblate in overall shape when furnished with lids.

As regards use context, the Utti Batue excavation sampled household debris (Bulbeck *et al.*, 2006), the Pattimang Tua and Pinanto excavations yielded some iron-working as well as domestic debris (Bulbeck and Caldwell, 2000), while Bola Merajae is poorly understood. Certainly, use of the vessels for cooking or holding heavy contents (such as water) can be ruled out given the unsuitability of the fabric for heavy-duty tasks, and the persistent association of soft pottery shards with shards from stronger, more serviceable vessels. The soft pottery vessels could conceivably have performed a ritual function, but this would imply that Bola Merajae was essentially a ritual site, based on its dominance of soft pottery (Table 1.2). It would be more reasonable to assume that the soft pottery at Bola Merajae probably had a wider array of functions than at the other three manufacturing sites.[4] Until further analysis may correct this impression, I infer that soft pottery had a domestic function. The wide-mouthed jars

would have been ideal for serving food (e.g., fruits, sago gruel, or cooked rice) to groups of diners; the contents would have been held secure by the inverted rim, while the wide aperture would have assisted serving or taking individual portions.

Finally, where do the origins of Luwu's soft pottery lie? Evidently, not in the Bugis heartland of the Cenrana Valley, to the southwest, which Bulbeck and Caldwell (2000) argue to have been the source of the Malangke Bugis immigrants. Over 5.5 kilograms of pottery dating to between the fourteenth and seventeenth centuries have been studied from the Bugis palace centre of Allang-kanangnge ri Latanete (Figure 1.1), which in many ways was Malangke's southern counterpart, and only a single shard of soft pottery has been identified (Bulbeck and Hakim, 2005). The most plausible source is Bola Merajae. The first millennium levels in Test Pit 1 yielded a small collection (19.7 grams) of shards very different from any contemporary pottery at Katue. The fabric (of Munsell brown coloration) resembles soft pottery in appearing porous and sandy, and speckled with gleaming inclusions (mica?), while the pottery is soft and low-fired. Indeed, soft pottery would appear to have been a specialty of the Bola Merajae potters, based on its prehistoric occurrence at the site, and the dominance of Soft Orange/Pink in the pre-Islamic levels. Either the Malangke and Pinanto potters imitated soft pottery vessels they had acquired from Bola Merajae, or Bola Merajae potters migrated to work in these population centres in Luwu's south. The latter development seems more likely, at least for Malangke, whose impressive population growth during the pre-Islamic period was associated with a multi-ethnic composition (Bulbeck and Caldwell, 2000).

A DARWINIAN EXPLANATION FOR THE RISE AND DECLINE OF LUWU'S SOFT POTTERY

As discussed above, market mechanisms are a suitable vehicle for natural selection in the case of artefacts where few consumers are also producers. The spectacular growth of Malangke's population, from an archaeolo-

---

[4] This observation holds even though the fragility of the soft pottery would lead to its over-representation in an assemblage of shards compared to the proportion of vessels in a household, at any time, that would have been soft pottery vessels.

gically invisible presence in the thirteenth century to over 10.000 in the sixteenth century, would have created a burgeoning demand for domestic pottery. Whether or not the Bugis settlers in Luwu had brought potters with them, there would have been considerable scope for diverse potters to ply their trade at Malangke, and other population centres (such as Pinanto) whose fortunes were tied to Malangke's. The producers of soft pottery appear to have specialized in making food holding and distribution vessels, which presumably replaced containers of metal, wood, or more durable earthenware previously used for that purpose. A chief advantage of the soft pottery is the basic nature of the technology. All that was required was suitable clay, not of high potting quality (hence, widely available), with a minimum of added temper and modest firing requirements. Soft pottery was evidently not only cheap but also fragile, further stoking its demand (cf. Arnold, 1985: 152-153).

The use of soft pottery evidently declined as the centuries passed (Table 1.2). Despite any attempts to improve the pottery's appearance with brownish slips or, in the case of Utti Batue, a whitish appearance, the misshapen appearance of the vessels would have been unmistakable.[5] Soft pottery, despite its cheapness, would have always been vulnerable to competition from other containers that looked more regular and aesthetic, and lasted longer, particularly in the well-off communities that evidently flourished at Malangke and Pinanto. This suggestion can be tested archaeologically based on the prediction that serving vessels of superior ceramic quality should have increased at Malangke and Pinanto concomitantly with the decline in soft pottery.

In summary, the unusual attributes of Luwu's soft pottery, and its chronological association with the period when Luwu's capital was based at Malangke, might entice explanations that appeal to introduced technology, advanced craftwork, or even ideological connotations. Analysis within a Darwinian framework, however, suggests a more prosaic explanation. Soft pottery was cheap and simple earthenware that filled a temporary niche in local society's demand at a time when population was burgeoning. This explanation is not only parsimonious but also hopefully inclusive, in the sense described in my Introduction. Attention has been paid to the Bugis pre-Islamic expansion into Luwu and the incorporation of an originally pre-Bugis technology into the Bugis-ruled economy. Inexact transmission of the technology is clear from the differences between the four known manufacturing sites in terms of their soft pottery, and adaptation to the environment is implied by the (suspected) use of local clay sources at these four sites. Now that a useful explanation for Luwu's enigmatic soft

pottery is available, it will be possible to incorporate it into a scientific explanation of social change, more generally, in pre-Islamic Luwu.

Acknowledgements

The "Origin of Complex Society in South Sulawesi" (OXIS) project was funded by the Australian Research Council, with supplementary support from the Wenner-Gren Foundation for Anthropological Research, the Australian National University, the University of Hull, and the Australian Institute of Nuclear Science and Engineering. The Australia-Indonesia Council funded my two trips to Makassar, South Sulawesi, to record the Luwu earthenware. Budianto Hakim of the Makassar Archaeology Office assisted in recording the earthenware. My attendance at the UISPP XV Congress was supported by a conference travel grant from the Faculty of Arts at the Australian National University.

References

ARNOLD, D.E. (1985) – *Ceramic Theory and Cultural Process*. Cambridge: Cambridge University Press. 237 p.

BOYD, R; RICHERSON, P.J. (1985) – *Culture and the Evolutionary Process*. Chicago: University of Chicago Press. 311 p.

BOYD, R; RICHERSON, P.J. (2005) – Rationality, imitation and tradition. In *The Origin and Evolution of Cultures*. Oxford: Oxford University Press. p. 379-396.

BULBECK, D. (2000) – Economy, military and ideology in pre-Islamic Luwu, South Sulawesi, Indonesia. *Australasian Historical Archaeology*. Sydney. S. 18, p. 3-15.

BULBECK, D. (In prep) – Early Metal Phase trade and craft in the Bone Gulf, Luwu, South Sulawesi, Indonesia.

BULBECK, D.; CALDWELL, I. (2000) – *Land of Iron: The Historical Archaeology of Luwu and the Cenrana Valley*. Hull: University of Hull. 141 p.

BULBECK, D.; HAKIM, B. (2005) – The earthenware from Allangkanangnge ri Latanete excavated in 1999. Accessible at WWW: URL: http://arts.anu.edu.au/bullda/Sarepao_pottery.html.

BULBECK, F.D.; BOWDERY, D.; FIELD, J.; PRASETYO, B. (2006) – The palace centre of sago city: Utti Batue site, Luwu, South Sulawesi, Indonesia. In Lilley M.; Ellis S., eds. lits. – *Wetland Archaeology & Environments: Regional Issues, Global Perspectives*. Oxford: Oxbow Books. p. 119-140.

CAMERON, D.W.; GROVES, C.P. (2004) – *Bones, Stones and Molecules: "Out of Africa" and Human*

---

[5] Given the qualities of soft pottery, the vessels could have lost their shape during use, and certain features such as rim grooves could be use-wear marks, so some of the shards' strange attributes may well reflect a different appearance of the worn-out wares compared to new vessels. Even then, however, it would hardly be an advertisement for a pot for it to sag and disfigure during use.

Origins. Burlington, MA: Elsevier Academic Press. 402 p.

GRIMES, C.E; GRIMES, B.D. (1987) – *Languages of South Sulawesi*. Canberra: The Australian National University. 208 p. Pacific Linguistics Series D – No. 78.

HAUSMAN, D.M. (1998) – *Causal Asymmetries*. Cambridge: Cambridge University Press. 300 p.

LAMPERT, C.D. (2003) – The Characterisation and Radiocarbon Dating of Archaeological Resins on Southeast Asian Ceramics. Bradford: University of Bradford Department of Archaeological Sciences. 350 p. PhD thesis.

LIEBERMAN, P. (2006) – *Toward an Evolutionary Biology of Language*. Cambridge, MA: Belknap Press. 427 p.

ODLING-SMEE, F.J.; FELDMAN, M.W.; LALAND, K.N. (2003) – *Niche Construction: The neglected Process in Evolution*. Princeton: Princeton University Press. 395 p.

PELRAS, C. (1996) – *The Bugis*. Oxford: Basil Blackwell. 353 p.

RENFREW, C.; BAHN, P. (2000) – *Archaeology: Theories Methods and Practice*. Third edition. London: Thames & Hudson. 640 p.

RINDOS, D. (1986) – The evolution of the capacity for culture: sociobiology, structuralism, and cultural selectionism. *Current Anthropology*, S. 27, p. 315-333.

THOMAS, D.H. (1972) – A computer simulation model of Great Basin Shoshone subsistence and settlement patterns. In Clarke, D.L., ed. lit., *Models in Archaeology*. London: Methuen. p. 671-704.

WHITTLE, A.W.R. (2003) – *The Archaeology of People: Dimensions of Neolithic Life*. New York: Routledge. 199 p.

WRIGHT, S. (1968 – 1978) – *Evolution and the Genetics of Populations*. Chicago: University of Chicago Press. 4 volumes

TEMPORAL TRENDS IN THE MORPHOMETRIC VARIATION OF THE LITHIC PROJECTILE POINTS DURING THE MIDDLE HOLOCENE OF SOUTHERN ANDES (PUNA REGION) – A COEVOLUTIONARY APPROACH

Marcelo CARDILLO

Departamento de Investigaciones Prehistóricas y Arqueológicas (DIPA-IMHICIHU) CONICET. Saavedra 15. 5to piso
marcelo.cardillo@gmail.com

Abstract: The focus of this paper is to discuss theoretical and methodological issues regarding the study of artifact's morphometric variation, which to some extent can be considered as an outcome of three interrelated dimensions: a) prior history (heredability of a particular shape), b) adaptation to proximate requirements and c) design constraints (architectural restrictions). The methodology proposed here, is based on the application of metrical and non metrical techniques, as geometric morphometrics. The last one, and the principal focus of this paper, allows us to perform shape and form analysis, treating these two variables independently one from the other. Based from the analysis of a small sample of middle Holocene projectiles lithiqués from the South Andes (Puna), we discuss here the potentials and limitations of the phylogenetic analysis over morphometric data. The results of this analysis suggest that while some basic traits of the form and shape have changed in different ways through time, other have tended to endure.
Keywords: Canalization Morphometric phylogenies design evolution

Resumé: Le objectif de ce papier doit discuter des aspects théoriques et méthodologiques quant à l'étude de la variation morphometric d'objet fabriqué, qui peut être considéré (dans une certaine mesure) comme un résultat de trois dimensions corrélatives: a) l'histoire préalable (heredité d'une forme particulière), b) l'adaptation aux exigences immédiates et c) aux contraintes de design (les restrictions architectoniques). La méthodologie proposée ici, est basée sur l'application de techniques métriques et non métriques, comme morphométrie géométrique. Le dernier et le principal foyer de ce papier, nous permettent d'effectuer analyse de forme et de taille, en traitant ces deux variables de façon indépendante un de l'autre. De l'analyse d'un petit échantillon de pointes de projectil du Holocène moyen dans les Andes Sud (Puna), nous discutons ici les potentiels et les restrictions de l'analyse phylogenetic sur les données morphometric. Les résultats de cette analyse suggèrent que pendant que quelques traits fondamentaux de la forme et de la forme ont changé différemment avec le temps, d'autre ont eu tendance à endurer.
Mots clés: La Canalisation Morphométrie phylogenies conçoit l'évolution

INTRODUCTION

In this work, I propose that a more powerful analysis of archaeological variation would result considering two dimensions of artifact morphologies: form and shape (Bookstein 1989). The definition of shape follows Kendall (1977) who identifies it as "…all the properties of a configuration of points those themselves are not altered for effects of size, position and orientation". Also, I use the term disparity (Zelditch *et al.* 2004) referring to morphological diversity. In a set of previous works (Cardillo 2002, 2004, 2005a y b), I explored the existing relations among form shape of lithic artifacts, especially projectile points. These analyses were framed within the research carried out by Hernán Muscio (Muscio 1996, 2004) about the evolution of human populations during the Middle and Late Holocene in the puna of Argentina. For this end, a theoretical model for the analysis of the observed variation in projectile points was proposed (Cardillo 2002, 2004). The goal of my previous work was to integrate the formal dimension (architectural constraints) to the study of projectile point variation. This paper also explore the historical and functional dimension through the concept developed by Gould (1989, 2002) of "adaptive triangle", to explain the evolution of adaptive morphologies (like projectile points). This model articulate three causal factors that

conforms a triangle in a theoretical space (Gould 1989, 2002 also see Cubo 2004). These factors are the structural dimension (geometry of the artifact, lithic raw materials), adaptational dimension (functional factors that arise by natural and cultural selection) and historical factors (phylogenetic patterns of trait variation and covariation). The different dimensions are interrelated an can be viewed like a frame to explain different evolutionary process. The relationship among the mentioned variables a dynamic and historic, so that they should be addressed in the ecological and temporal framework pertinent for each case, as in our case is the Middle and Late Holocene in the Puna of Salta, Argentina (see below). In these analyses I used tools of traditional and geometric morphometrics (Rohlf 1990) using variables as the long, wide, thickness and weight as well as different indexes (Cardillo 2004). Geometric morphometrics analysis follows standard methodological procedures treating landmarks and contours (Bookstein 1991, Rholf 1990) that were developed mainly by Leslie Marcus, James Rohlf, Fred Bookstein and others (Marcus *et al.* 1993, Adams 2004). In morphometrics a landmark is a point located by anatomical criteria and consistent with biological homology (Bookstein 1991) in archaeology this point, can be the apex of a projectile point or other fixed point in morphology. In the case of no clear homology between points, we use the term semilandmarks (Bookstein 1996/1997). These points can be the endpoints of a minimum or maximum diameter of a form. According to this, we use semilandmarks to describe the contour of projectile points. In this case, the entire outline could be treated as an hypothetical homologous unit.

The use of these techniques suggests the existence of different patterns of morphological change inside the studied sequence, and a null relation between life history and morphological variability, (basically between shape and reactivation); and a significant linear relation between size and shape of the instruments (allometry, by geometric canalization) (Cardillo 2004). Also, temporal differences between morphotypes suggest changes in mean shape though time. Within a coevolutionary framework, (Durham 1991) I propose that variation in performance requirements, design constraints and cultural transmission patterns[1] related with shift in human adaptations (to a extended discussion of this issue see Lopez in this volume) are important factors to explain evolutionary process in the middle and late holocene of Argentinean Puna.

Using the adaptive triangle as a conceptual frame, this work explores how the relation between design options on performance requirements and architectural constraints, could have been driving variability in large time scales (Cardillo 2005a and b). Following Hughes (1996) and Ratto (2003), I suggest the existence of less morphological variation in great size lanceolate designs than in more oval ones. This fact are in part related with performance requirements for different functions, not necessarily restricted to serve as throwing weapons, but sometimes also serving as hand throwing spears (Ratto 2003, Martínez 1999) as well as to geometrical canalization. To explore this historical patterns I will use teorethical morphospace[2] models (see below). Expectations are that the relation between functional, structural, and historical dimensions produced heterogenous morphospace occupation.

However I realize other factors that the averaged sample of morphologies may result in false patterns (Bush et al. 2002). Rare morphologies could be absent or even worst, inflating disparity much than in reality. Both aspects may result in spurious patterns leading us to make false inferences, for which there are kept in mind when considering results.

This work proceed in three basic steps:

1. The application of distance-based phylogenetic methods (Neighbor-joining) over shape coordinates, as a way to explore historical patterns in continuous morphological characters.

2. The application of partial least square procedure and the Mantel test to explore patterns of variation and covariation between morphotypes.

3. The representation of the observed patterns in a theoretical morphospace that comprise the different dimensions of the adaptive triangle.

STUDY AREA

The sample was obtained from surface and sub-surface contexts in Ramadas site, located in San Antonio de los Cobres valley, Puna of Argentina. The temporal span of the occupation of Ramadas is between 6000 bp and 4000 bp. The technological sequence of this site is similar to others sequences recorded in the dry and salty Puna, in sites as those from Quebrada Seca e Inca Cueva 4 (Martínez 1999)

MATERIALS AND METHODS

I selected a sample of ten morphotypes (or morphological variants), nine of them correspond to the samples collected in the studied zone but, one of these (the selected outgroup -OG- that dates from 6000 bp) (fig 2.1) proceeds from the site Quebrada Seca 3 excavated by Carlos Aschero (Aschero 1987, Aschero et al. 1993-1994). These types were taken from a larger collection after a prior morphometric analysis that allowed to discriminate some basic trends in form and shape. Also, I defined some classes based on the presence of particular discrete traits like a denticulate limb or shoulders. The selected morphotypes are well represented in the archaeological sequences of very different sites with also good chronological information (see Ratto 2003, Martínez 1997). I used this information to build temporal blocks and in this way explore the observed patterns of disparity.

MORPHOMETRIC ANALYSIS

Morphometric analysis follows the methodology proposed by Rohlf (1990) and Bookstein (1991) among others. Starting with a set of landmarks placed along the outline of previous digitized images, using the *tps dig* program (Rohlf 2004). Applying this methodology, I chose two landmarks and 23 semilandmarks in order to describe the basic geometry of the projectile points. I used only half contour to avoid the asymmetry and redundancy (for copying two similar sides). The complete form was only used for better graphic display of hypothetical ancestors and observed tendencies in morphological change over the tree. For this particular case I used a total of 140 points (see below).

---

[1] Cultural transmission play a significant role in two teorethical dimensions of this triangle: functional (adaptive learning) and historical (trial and error learning, biases in trait adoption and transmission (Cavalli-Sforza y Feldman 1981, Boyd y Richerson 1985)).

[2] Morphospace or morphological space (McGhee 1998) is a geometric character space with *n* dimensions, inside which is possible to study the relationships between a set of forms observed empirically and those that are unobserved (or not exist in a sample) but are theoretically possible.

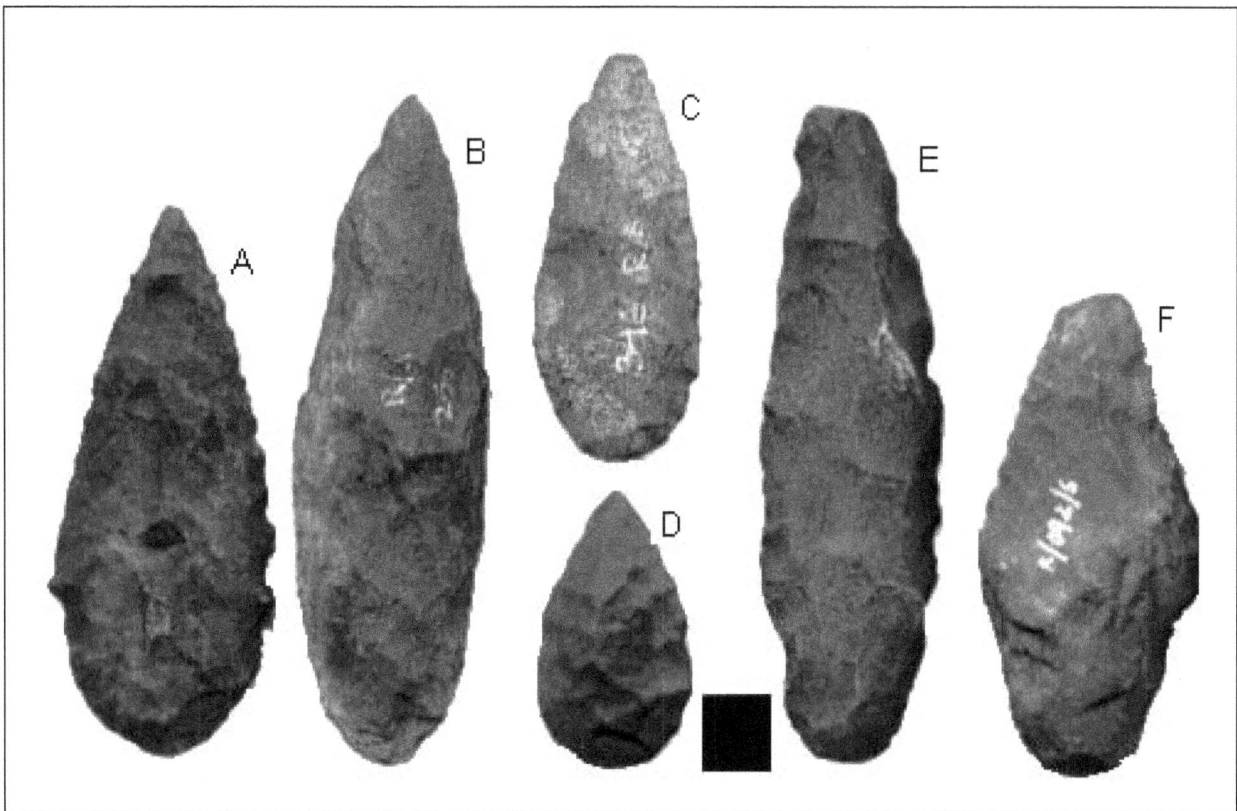

Fig. 2.1. Different morphotypes: A 5000 B.P., B 6000 B.P., C 4000 B.P., D 4000 B.P., E 6000 A.P, F 5000 B.P.

PHYLOGENETIC ANALYSIS

The neighbor-Joining method was proposed by Saitou and Nei (NJ) to analyze distance data (Saitou and Nei 1987). This procedure allows the construction of phenetic trees from continuous data like morphological multivariate data. This is a very especial kind of data because each vector of shape is non-independent of each other (see discussion in Monteiro 2000). Only the sum of shape vectors makes sense for the analysis. This fact precludes the use of discrete points (like binary characters commonly used in maximum parsimony methods) or other arbitrary cut-offs of the data. Saitu and Nei, among others (Saitou and Nei 1987, Kim *et al.* 1993) showed convincingly through simulation and real data that NJ produces similar results than parsimony, maximum likewood and others methods, performing better than the first in cases of disparate rates of change (Saitou and Nei 1987). Although, with this method the change in character states can't be rebuilt (each landmark contributes equally to distance between cases), being possible to analyze the congruence with ordinary resampling routines. The resulting NJ tree is an unrooted tree, but can be rooted equally like in maximum parsimony based methods. Neighbor-Joining produces additive trees; this means an assumption of a constant evolutionary rate. For this last reason, longest branches imply more accumulated change; hence, NJ method is accurate building hypothesis over rates of change in morphometric traits.

After obtaining raw data, all individual specimens were subjected to generalized Procrustes analysis. The generalized Procrustes analysis is a least-squares (LS) superimposition method that translates, rotates, and scales the landmarks for each individual (Rohlf and Corti 2000). The distance matrix was Euclidean distances among types based on Procrustes coordinates for each morphotype. However differences between curve space of Procrustes coordinates and linear Euclidian space must be controlled to avoid errors of measurement. Error was measured with a correlation between both spaces using *tpssmall* (Rholf 2003b) program (slope= 0.999454, correlation $r=1$)

NJ tree was made with Past 1.56b. A resampling method (bootstrap) was used to analyze the support for each branch. Bootstrap support was determined by randomly sampling (1000 times) 25 landmarks from Procrustes matrix and calculating Euclidian distances for each case and the corresponding tree, using the Neighbor-Joining method. Values for nodes: 2=100, 3=22, 4=68, 5=65, 6=89, 7=56, 8=70, 9=100 (figure 2.2).

RECONSTRUCTION OF INTERNAL NODES OVER RESULTING TOPOLOGY

As S J Gould (1991) stated, while cladistics is an efficient method for branching order reconstruction in a genealogy, disparity is a phenetic issue. Thus a multivariate approach to variation is needed to make account to the relation

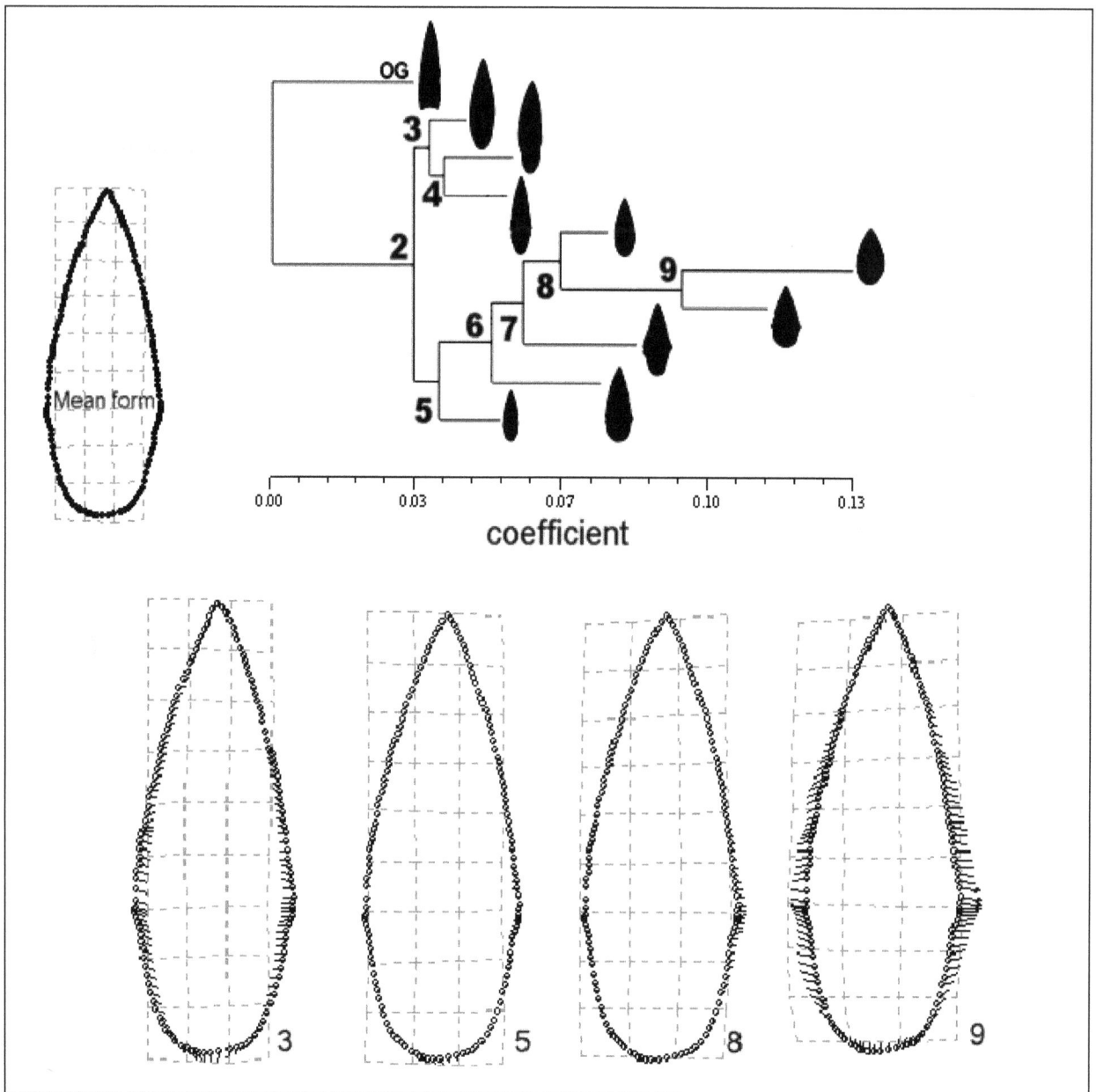

Fig. 2.2. Upper section of the figure: Resulting NJ tree over 25 points (2 landmarks and 23 semilandmarks) points were previously aligned with generalized Procrustes superimposition procedure which was the base for the Euclidian distance matrix among cases. Below: expected shape to four nodes using a least square superposition criterion between the observed phylogeny and a very densely sampling outline (140 points, two landmarks and 138 semilandmarks). Black circles show the mean shape, while vectors and grills signal the mount of deformation from the mean shape

between both processes. But to assess the relation of morphological distances with the branching structure observed in a clade is another problem. However, it is possible to use partial least square approximation (Rohlf 1998) to explore the covariance between morphology and a clade. With this goal, I used the *tpstree* program, developed by Rohlf (2003c). This program computes the orthogonal least-squares Procrustes average configuration of landmarks using generalized Procrustes analysis (Rohlf and Slice 1990). The consensus configuration is rotated to find the axis of maximum covariation between a multi-

variate shape and the morphotypes used in any observed additive tree. In this way the program fits shape data to a tree in the sense of estimating the shape of the nodes or HTUs (Hypothetical Taxonomic Units) in the tree (Rohlf 2003c). Therefore, least squares allow us to explore the changes in shape corresponding to different positions in the tree. Note than I use the mean shape and not the Outgroup to PLS analysis. This is because the increase of the variability observed in the tree is a function of polarization due the outgroup selection, resulting in equifinality.

DISCUSSION

The NJ tree suggests an increment of distances correlated with an allometric relationship between shape and size (see figure 2.2). The bootstrap result shows a relative good support for the majority of the clades. A major issue is that discrete traits along outline like denticulate, or serrated edge appears having not-weight in the analysis (appears as homoplasious features). It seems obvious that when neighbors in the tree are computed, this small scale variation contribute less to the global variation than the compression-dilatation of the shape (Cardillo 2006 MS). This suggests that the method performs better with global scale modifications like allometric variation. It is possible to weight small scale variation using other superimposition methods. But the potential outcomes of this procedure over final results is not easy to evaluate yet , because it is not clear which features have a potential phylogenetic signal, and extra assumptions need to be accompanied by more research.

The estimated HTUs using PLS procedure shows the hypothetical ancestral forms from an empirical distribution. The maximum covariance between shape and NJ tree depicts an increase of morphological variation correlated with an increasing phylogenetic distance. The amount of change is represented as vectors and deformation grids over the average configuration. These vectors show a gradual expansion in the mean sector and a minor change in the base and the tip. This minor relative change might respond to functional aspects, like shaft requirements regarding the base and the penetration capacity of the projectile point. The increase of the change observed at HTUs is evident since the 8th node (fig 2.1) is a sinapomophy or derivative with respect to clades 2, 3, 7 (shapes chronologically located in 6000-5000). The results of the phylogenetic analysis also shows a coherent pattern with the chronological data. A Mantel test between temporary blocks defined early in this paper (6000, 5000, 4000 see above) and the procrustes coordinates shows a moderate correlation between morphological distances and time (R=0.48, p<0.002). This results must be taken with caution bearing in mind possible biases related with sample size and segmentation criteria, but suggest a possible relation between disparity and time.

As a whole, the results suggest a tendency not explained by chance. This increase of change would be major towards 4000 B.P. where there is well based archaeological evidence of changes in human subsistence strategies (Yacobaccio 1994, Muscio 1996, Ratto 2003) mainly the reduction of the residential mobility and the beginning of the camelids domestication practices. This might have changed the role of the extractive strategies affecting the requests of performance of hunt technologies, as the projectile points. In previous papers (Cardillo 2002, 2005a y b), I proposed that subsistence change could relaxed the existing pressures of selection on extractive technology allowing the emergence of new

morphotypes due to an increase of neutral variation. But the actual results can be also a by-product of different process like competition at the design level (that not always results in an optimal solution due to canalization) related to competition for resources at human population level as Lopez (in this volume) suggest. Also, the experimental work of Martinez (1997) support the idea of different performance properties (strengh to impact, cutting and penetration capacity) of some of this morphologies.

From my point of view, cultural innovation, and morphological canalization under structural constraints played an important role in design evolution with adaptive cultural transmission (and maybe natural selection in the long run) acting as functional process. The temporal persistence of this artifactual design plan also suggests a phylogenetic signal, probably as a result of the vertical transmission within the same cultural population.

To address this question, an increase of sample size and better chronological information are required, as well as inter and intra artifact group variation analyses. However, empirical morphospaces are related to sample dimension and composition (McGhee 1999). The empirical space of forms will always be dependent on the units that compose it (from which it was inferred). In contrast, theoretical morphospaces are determined by geometric models, not by sample measurements (McGhee 1999). To display morphological change and design constraints through time, it is possible to generate a theoretical space of shape, from a limited number of variables. In this case, that space is a combination of different coordinates along two dimensional axes. In this axes hypothetical combinations of variables show the whole potential change in shape in relation to size. In this way, it is possible to add empirical cases to this theoretical space without modifying its basic configuration. This allows us to compare different sets too. The parameters of the model was the height/wide ratio and the weight, with this parameters I generate a two-dimensional space that shows the whole possible changes in morphology and size. All possible combinations of this variables was made in an equally spaced intervals from 0.1 to 1 (height/wide index values) and from one to seven grams. After this, I plotted all the observed and empirical cases represented as a frequency distribution inside the theoretical space. Lanceolate points from other site (Inca Cueva7) were included too. Based on the observed pattern we propose a general model of design evolution within the theoretical morphospace (figure 2.3)

The empirical units show an anisomorphic pattern of distribution within the theoretical morphospace probably related to architectural constraints. This distribution also shows an allometric relation between size and shape, (more weigh projectile points have low index values and viceversa). It is important to note that changes in size may produce sorting in shape and a possible range of sorted neutral variation or perhaps, suboptimal designs. Early in

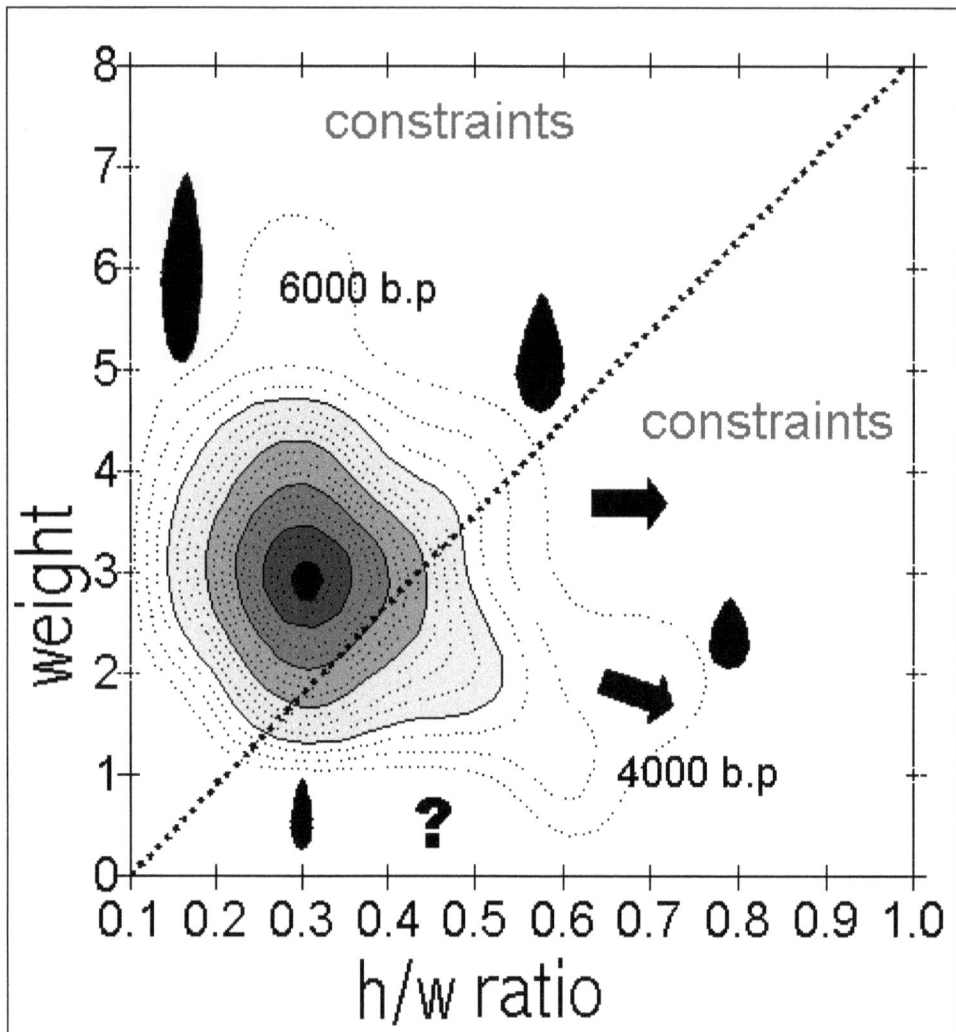

Fig. 2.3. A morphospace evolution model for the 6000-4000 bp time span. The arrows show the hypothetical direction of change thought time over the frequency distribution of projectile points in theoretical morphospace. Extreme (observed) morphologies are also depicted in the boundaries of the empirical distribution. Interrogation sign (?) suggest a maybe available -but actually- non-occupied morphospace. Relative form elongation is represented here like an allometric relation h/w (height/wide ratio). Dashed line depicts the isometric grow vector
* Contours measure the increase of density (in grayscale) per unit area of the plot

this paper I propose (following Gould 1989, 2002) that this pattern only can be fully understood at the light of three statements that conforms an "adaptive triangle": functional vertex (active adaptation thourgh natural and cultural selection) historical vertex (constraints of phylogeny) and formal vertex (rules of structure). At this light, coevolution of projectile point morphology and human adaptation was not only related to functional factors (adaptive goals); structural elements (geometrical constraints over design options) and phylogeny (heredity patterns and prior history) also affected the evolution of design thourgh canalization.

Finally, disparity comes out from the interaction of this three elements. This morphospace model also shows that not all the potential morphological variation is present. This pattern suggesting canalization and selection toward

adaptive (but in some cases suboptimal) designs. However, the relation between potential and realized variation (as we seen in the morphospace model above) need to be explore yet.

All of these results need to be studied more carefully. The application of other phylogenetic methods like additive trees with reticulations and more complex form spaces building are the next steps. A comparative approach (Mace and Pagel 1994, Mace et al. 2003) using different information seems to be necessary too.

CONCLUSIONS

The relation between functional and structural factor may help us to explain the evolution of projectile point's designs during the Middle-Late Holocene in the puna region of Argentina. The branching order of the Neighbor-joining tree and the patterns of variation and covariation between morphological and metrical traits suggest a complex history. To explain this, a directed or anisotrophic pattern of cladogenesis is proposed. Finally, the evolutionary process inferred here, appears like a compromise between different dimensions, history, functional requirements and design constraints.

Acknowledgements

I´m specially thank Hernan Muscio and Gabriel López to invite me at this symposium. I would like to express my gratitude to Luis Alberto Borrero and Hernán Muscio for the very useful comments to previous version of this paper. Much of this work would not to be possible without the help of my teachers in the postgraduate course in Theoretical Systematics, Dra Viviana Confaloneri, Lic Shirley Spert, Lic Santiago Catalano and Lic Marcela Rodriguero to our disposition to listen, share knowledge and discuss much of my ideas about cultural phylogenies. To Lic Federico Giri who help me in process morphometric data. At last I'm very grateful with all of the scholars that share with me invaluable bibliography, Ivan Perez, James Rohlf, Miriam Zeldrich, Ignacio García and Takao Ubukata. This work is dedicated to amazing brother Pablo.

References

ADAMS, D.C., F.J. ROHLF, and D.E. SLICE (2004). Geometric Morphometrics: Ten Years of Progress Following the 'Revolution'. Italian Journal of Zoology. 71, p. 5-16.

ASCHERO, C.A. (1988). El sitio ICC4: un asentamiento precerámico en la Quebrado de Inca Cueva (Jujuy, Argentina) Estudios Atacameños 7. Universidad Católica del Norte. San Pedro de Atacama. Chile.

ASCHERO, C.A. MANZI L.M and A.G. GOMEZ (1993-1994). producción lítica y uso del espacio en el nivel 2b4 de Quebrada seca 3. Relaciones 19. Sociedad Argentina de Antropología. Buenos Aires.

BOOKSTEIN, F.L (1989). "Size and Shape" a comment on semantics. Syst.Zool 38, p.173-180.

BOOKSTEIN, F.L (1991). Morphometrics tools for landmarks data. Geometry and biology. Cambridge University Press. New York.

BOOKSTEIN, F.L. (1996/7). Landmarks methods for form withoutt landmarks: morphometrics of group differences in outline shape. Medical Image Analysis,. Oxford University press. 1:3, p. 225-243.

BOYD, R. Y P.J. RICHERSON (1985). Culture and the Evolutionary Process. University of Chicago Press. Chicago. IL.

BUSH ANDREW, M; POWELL MATTHEW G; ARNOLD WILLIAM S; BERT THERESA M and GWEN M DALEY. (2002). Time Averaging, evolution and morphologic variation. Paleobiology 28:1, p. 9-25.

CARDILLO, M. (2002) Transmisión Cultural y Persistencia Diferencial de Rasgos. Un Modelo para el Estudio de la Variación Morfológica de las Puntas de Proyectil Lanceoladas de San Antonio de los Cobres, Provincia de Salta, Argentina. Perspectivas Integradoras entre Arqueología y Evolución. Teoría Método y Casos de Aplicación. Editado por Martínez GA y Lanata JL. INCUAPA. serie teórica 1, p. 97-199.

CARDILLO, M. (2004). Arqueología y procesos transmisión cultural. Una aproximación teórico-metodológica. Tesis de Licenciatura. Facultad de Filosofía y Letras (UBA). Argentina. Manuscrito.

CARDILLO, M. (2005a). Explorando la variación en las morfologías líticas a partir de la técnicas de análisis de contornos. El caso de las puntas de proyectil del holoceno medio-tardío de la Puna de Salta (San Antonio de los Cobres, Argentina). Werken 7, p. 77-88.

CARDILLO, M. (2005b). Exploración de patrones de variación morfológica en los artefactos líticos. Una aproximación desde la morfometría geométrica. Submitted to Complutum.

CARDILLO, M. (2006). Contribution to the application of morphometric and phylogenetic based methods in archaeology. The use of experimental morphometric sequences. MS.

CAVALLI-SFORZA, L.L. and M.W. FELDMAN (1981). Cultural Transmission and Evolution: A Quantitative Approach. Princeton University Press, Princeton.

CIAMPAGLIO, C., M. KEMP and D. MCSHEA (2001). Detecting changes in morphospace occupation patterns in the fossil record: characterizationand analysis of measures of disparity. Paleobiology 27, p. 695–715.

CUBO, JORGE (2004). Patter and process in constructional morphology. Evolution & Development 6:3, p. 133-135.

DURHAM, W.H. (1991). Coevolution: Genes Culture and Human Diversity. Stanford University Press. Stanford, California.

GOULD; S.J. (1989). A developmental constraint in Cerion, with comments on the definition and interpretation of constraint in evolution. Paleobiology 43:3, p. 516-539.

GOULD, S.J. (1991). The Disparity of Burgess Shale arthropod fauna and the limits of cladistics analysis. Why must strive to qualify morphospace. Paleobiology 17, p. 411-423.

GOULD, S.J. (2002). La estructura de la teoría evolutiva. Tusquets. Barcelona. España.

HUGHES, S.L. (1996). Getting to the Point: Evolutionary Change in Prehistoric Weaponry. Journal of Archaeological Method and Theory. Vol 5 (4), p. 345-408.

KENDALL, D.G. (1977). The diffusion of shape. Advances in applied probability 9, p. 428-430.

KIM, J., F.J. ROHLF, and R.R. SOKAL (1993). The accuracy of phylogenetic estimation using the neighbor-joining method. Evolution, 47, p. 471-486.

MACE, R. and M. PAGEL (1994) Comparative method in Anthropology. Current Anthropology. 35, p. 549-564.

MACE, R., F. JORDAN and C. HOLDEN (2003). Testing evolutionary hipotesis about human biological adaptation using cross-cultural comparison. Comparative Biochemistry and physiology. Part A 136, p. 85-94.

MARCUS, L.F., E. BELLO, and A. GARCÍA-VALDE-CASAS (1993). Contributions to morphometrics. Editado por Marcus, L.F., E. Bello, y A. García-Valdecasas. Monografias del Museo Nacional de Ciencias Naturales 8, Madrid.

MARTÍNEZ, J.G. (1999). Puntas de proyectil, diseño y materias primas. En los tres reinos: prácticas de recolección en el Cono Sur de América. Editado por Aschero C.A, Korstanje M.A y Vuoto P.M. Magna Publicaciones, p. 61-69. Universidad Nacional del Tucumán. Argentina.

MCGHEE, J.R., GEORGE R. (1999). Theoretical Morphology. The concept and its applications. Columbia University Press. NY.

MONTEIRO LEANDRO, R. (2000). Why Morphometrics is special: The Problem with Using Partial Warps as Characters for Phylogenetic Inference. Systematic Biology 49:4, p. 796-800.

MUSCIO, H.J. (1996). Poblamiento Humano y Evolución en la Puna Argentina. Desarrollo Teórico para la Arqueología del valle de San Antonio de los Cobres, Salta. Tesis de licenciatura FFyL.UBA. MS.

MUSCIO, J. (2004). Dinámica poblacional y evolución durante el período Agroalfarero Temprano en el Valle de San Antonio de los Cobres, Puna de Salta, Argentina. Tesis doctoral presentada a la Facultad de Filosofía y Letras, Universidad de Buenos Aires.

HAMMER, Ø., HARPER, D.A.T., and P.D. RYAN (2001). PAST: Paleontological Statistics Software Package for Education and Data Analysis. Palaeontologia Electronica 4 (1): 9. http://palaeo-electronica.org/2001_1/past/issue1_01.htm.

ROHLF, F.J. (1998). On applications of geometric morphometrics to studies of ontogeny and phylogeny. Systematic Biology, 47, p. 147-158.

ROHLF, F.J. and D. SLICE. (1990). Extensions of the Procrustes method for the optimal superimposition of landmarks. Systematic Zool., 39, p. 40-59.

ROHLF, F.J. (1990). Morphometrics. Ann. Rev. Ecology and Systematics, 12, p. 299-316.

ROHLF, F.J. (1990). Fitting curves to outlines. Rohlf, F.J. and F.L. Bookstein (eds.). Proc. Mich. Morphometrics Workshop. Univ. of Michigan Museum of Zoology (Special Publication no. 2), p.167-178.

ROHLF, F.J and M CORTI (2000). The use of partial least-squares to study covariation in shape. Systematic Biology 49.

ROHLF, F.J. (2003a). tpsSmall, version 1.20. Department of Ecology and Evolution, State University of New York at Stony Brook.

ROHLF, F.J. (2003b). tpsTree, fitting shapes to trees, version 1.18. Department of Ecology and Evolution, State University of New York at Stony Brook.

ROHLF, F.J. (2004). tpsDig version 1.40. Department of Ecology and Evolution, State University of New York at Stony Brook.

ROHLF, F.J. (2002). Geometric morphometrics in phylogeny. Forey, P. and N. Macleod (eds.) Morphology, shape and phylogenetics. Francis & Taylor: London, p. 175-193.

SAITOU NARUYA and MASATOSHI NEI (1987). The Neigbor-Joining Method: A New Metod for Reconstructing Pylogenetic Trees. Mol.Biol. Evol 4:4, p. 406-425.

RATTO, N. (2003). Estrategias de caza y propiedades del registro arqueológico en la Puna de Chaschuil (Dpto de Tinogasta, Catamarca, Argentina). Tesis de Doctorado. Facultad de Filosofía y Letras. Universidad de Buenos Aires.

YACCOBACCIO, H.D. (1994). Biomasa Animal y Consumo en el Pleistoceno-Holoceno Surandino: Arqueología 4.

ZELDITCH MIRIAM, L., DONALD L. SWIDERSKI; SHEETS DAVID H; FINK WILLIAM L. (2004). Geometric Morphometrics for Biologist. A Primer. Elsevier.Academic Press. NY.

INTERDEMIC SELECTION AND PHOENICIAN PRIESTHOOD – DARWINIAN REFLECTIONS ON THE ARCHAEOASTRONOMY OF SOUTHERN SPAIN

José Luis ESCACENA CARRASCO, Daniel García RIVERO

Departamento de Prehistoria y Arqueología. Facultad de Geografía e Historia
Universidad de Sevilla. C/ María de Padilla s/n, 41004, Seville, Spain
E-mail: escacena@us.es / garciarivero@us.es

Abstract: In their expansion towards the West, the Phoenicians put into practice the astronomical knowledge gained in their temples. As the holders of this scientific experience, the priests contributed to the colonial expansion as hypermutators of behaviour: first they acquired the celestial knowledge, then they applied it to nautical orientation. From an evolutive perspective, this implied a very adaptive mechanism for demographic growth. However, this system required the encoding of the positive memetic mutations in order to prevent them from transmitting to Greek seafarers that competed with the Phoenicians within the diaspora.
Key words: Evolutionary archaeology, Interdemic selection, Priesthood, Temples, Archaeoastronomy

Résumé: Lors de leur expansion vers l'Ouest, les Phéniciens ont mis en pratique les connaissances astronomiques acquits dans leurs temples. En tant que détenteurs de cette expérience scientifique, les prêtres ont contribué a l'expansion colonial comme hypermutateurs du comportement : en premier lieu ils acquéraient le savoir célestiel, puis ils l'appliquaient a l'orientation nautique. D'un point de vue évolutif, ceci impliquait un mécanisme très adaptatif pour la croissance démographique. Cependant, ce système demandait la codification des mutations mémétiques positives pour empêcher leur transmission aux navigateurs Grecques qui rivalisaient avec les Phéniciens dans la diaspora.
Mots clés: Archéologie évolutive, Sélection interdémique, Clergé, Temples, Archéoastronomie

THE THEORETICAL FRAMEWORK

Mainstream interpretation considers religion as a mechanism of reproduction for social structures and economic inequalities which, since the Neolithic, have shaped human groups. This reading however fails to explain why religious behaviours became generalized in all cultures if they only benefited restricted groups within each community: the elites. Moreover, because of the general rejection of biology by historians as a field that contributes towards historical research, specialists in humanities usually ignore the connections between faith and antistress mechanisms (Tobeña 2005: 212) or the immune system (Punset 2004: 17). This situation is exacerbated by the recent relationship between Darwinism and Archaeology. Indeed, this field has only recently begun its breakthrough (cf. Rindos 1984; 1988; Maschner 1996; Hart and Terrell 2002).

Religions collaborate in the reproduction of the social structures of which they are part, but this does not demonstrate that the beneficiaries of such mechanisms of replica are necessarily the highest groups on the social scale. If the mere existence of these elites is evidence of inequality, from a Darwinian perspective it can be argued that the internal hierarchization of a community is narrowly related to intergroup competition over resources. This evolutive pressure corresponds to the concept of "interdemic selection" put forward by V.C. Wynne Edwards (1963). This leads to accept that natural selection can in some cases act upon human groups as units of selection in such a way that the beliefs of each group becomes one of the factors that reinforces the ethnic boundaries between sympatric communities that

compete for the control of the environment. In most social animals, this phenomenon has caused a spontaneous bias towards intragroup stratification over a time span of millions of years, without there being any scientific reason for which to exclude *Homo sapiens sapiens* from this trend.

Following on from these premises, it is likely that the Phoenicians of the 1st Millennium BC – and not only their upper social classes- benefited from their national religion in their competition with other groups that followed their footsteps in their Mediterranean expansion, particularly the Greek community. The Canaanites tactics, obviously not exclusive to this population, were based on a design of demographic expansion by means of a colonisation planned in the temples, in which the maritime routes were plotted and the correct location of the new enclaves decided. In this matter, the knowledge of the sky of the clergy constituted a key tool, without which it would have been impossible to carry out this kind of astronomical navigation and the establishment of the closed commercial circuits between East and West. This role had not been assumed by the Canaanite priests prior to the 1st Millennium BC. We could thus be faced with an exaptation of the kind proposed by S.J. Gould and E.S. Vrba (1982). However, this term should not imply any special recognition of the originality of the changes. In fact, if the Darwinian concept of evolution denies the teleological dimension of the exaptation, nothing emerges for anything and everything fulfils previously another function. Indeed, any thing of nature, be it a somatic organ or a pattern of behaviour, would be an exaptation of a previous adaptation.

Studies of the Phoenician colonisation based on this perspective of the issue are completely inexistent. First, because it is widely considered by specialists in social sciences that the evolutive theory does not serve in the analysis of the more recent periods of human history, and second, because archaeologists have been reticent to interpret as cosmic symbols or as astronomical knowledge much of the documentation that they are faced with. Although Iberian Protohistory is now more open to accept the astral orientation of the sanctuaries and other cult structures, the Darwinian analysis of such questions has never been carried out. On the contrary, it has even occasionally been argued that Archaeology is a Lamarckian science (cf. Querol 2001: 35).

Since Darwin published his work on the origin of species, his theory was immediately applied to human evolution, thus leading to his later book on our ancestry. There was no reason for man to be an exception to the rule. The Darwinian approach has constituted since then a form of understanding designed to operate with individuals and groups, in both their somatic or physiological aspects and their forms of behaviour. The inclusion of behavioural aspects is the point of disagreement with all other theoretical positions in historical analysis. Evolutive Archaeology would suggest on the other hand that change through selection is of universal application, that it explains the past and the present, the somatic, the physiological and the behavioural, and that ultimately culture – and technology as a part of it- would evolve in the same way.

GENES *VERSUS* MEMES?

Recently among Darwinists, some degree of controversy has emerged over the question of whether genes are the basic motor of human evolution or whether they have been overtaken by memes (cf. Alexander 1994: 74; Blackmore 2000: 143-177). The Phoenician diaspora shows, none the less, a symbiotic cooperation between the two types of replicants. As any mutualism, the alliance was beneficial for both parties, thus contributing to the demographic expansion of their descendents (the Canaanites), first throughout the Mediterranean and later throughout the Atlantic. In any case, it appears that the mutations experienced in the centuries of this process of diffusion were never deep enough as to adapt the Semitic culture to ecosystems different to the subtropical conditions in which they originated. For this reason, the Phoenician dispersion, as many others migrations (Diamond 2001: 88-89), found ease in its horizontal expansion but was unable to move along the meridians, once out of the Mediterranean, further than the latitudes tolerated by the Mediterranean agriculture that constituted the basis of its subsistence: the Portuguese coast to the north and Morocco to the south.

The opposition between genes and memes as motors of change by means of their own mutations constitutes another trap in which non-epistemic values interfere in the scientific analysis. It stems from the influence on logical analysis of a dichotomic conception of man that divides the body (somatic part) from the soul (origin of behaviour). In contrast, for Darwinian analysis, that only aims to understand how life functions and not to show the direction in which it should be heading, it is only possible to think of the human individual in terms of a undividable self. From this point of view, natural selection would be unable to discriminate between body and behaviour, nor to act upon the former and not the latter, for the simple reason that there can be no body without behaviour nor vice versa. By means of the application of astronomical knowledge to nautical orientation, the genes of the Phoenician populations reached a previously unknown rate of expansion. Moreover, in every new colony, where a large number of new genetic replicas took place, the Phoenician memes also proliferated, reproducing their social organization, beliefs, eating habits, language, family system, political regimes, technology, etc.

THE PHOENICIAN CLERGY AND ASTRONOMY – A DARWINIAN ANALYSIS OF THE CANAANITE DIASPORA

Some of the keys of evolution are now being discovered in microbiology. The concept of the *individual* itself has even become questioned in microscopic life (Margulis 2003: 118). Symbiosis that exceeds more than mere mutualism, or beings that only prosper as collectivities, sheds reasonable doubt over what may be the minimal units of selection. In microscopic life, variation increases through contingent mutations but also through the horizontal exchange of genetic material. At first, this mechanism could resemble a Lamarckian evolution since these acquisitions are hereditary, however what this really creates is a fertile field upon which natural selection operates. The horizontal shift of genes, frequent among bacteria, thus offers a valuable model for the analysis of the similar human cultural transmission. Equally, as microbial life usually prospers in communities and the exchange of genetic material takes place to a greater extent within these communities, its study provides paradigmatic examples by which to explain human phenomena of group selection in which behaviours of restricted permeability of memetic information can be observed.

Interdemic selection appears when the members of a population contribute to the descendency of the group in a non random way, that depends upon their behaviour. Thus, Nature opts for the behaviours that favour demographic growth, choosing those that increase the offspring. However, for this selective mechanism to take place the level of variation between individuals of the same group must be less than that between the separate groups (Boyd and Silk 2001: 220). We can accept that this condition of memetic distance was fulfilled in the Western Mediterranean territories. Indeed, the existence of multicommunitarian situations in the Phoenician colonial provinces (in Tartessos for instance) is now considered the most likely setting. As a result, these social (and in this case cultural) groups offer themselves to natural selection as true "units of choice" that enter into evolutive competition with other "units" that represent the other choices. In our case, interdemic selection does not appear to have operated only upon conflicts of demographic growth between the resident population of each territory and the Semitic group, but also between the latter and the groups who would, from other parts of Eastern Mediterranean, later attempt to open routes towards the West: the Greeks.

Behavioural studies see religious conducts as an ideal field for evolutionary experimentation (Burkert 1996; Dennett 1998; Lincoln 1981). In this sense, it is usually the case that beliefs are more easily transmitted at a young age to the members of the same culture, than to adult members of a different culture. Despite this, the issue has been dealt with the other way round by most of the specialists that have studied the Semitic colonisation of Tartessos. With the exception of J. Alvar (1993) few have questioned the permeability of the indigenous populations faced with a foreign religious universe.

Religion fulfils several evolutive functions, some of which have been examined from a Darwinian perspective. However, for Darwinian analysis it is of less interest to question how and why religious conducts emerged, since they probably did so as a subproduct of symbolic thought. It is thus of greater value to understand why beliefs constitute today a common practice in all cultures. This in itself says a lot about its positive contribution to the reproduction of individuals and populations. Indeed, the optimism offered by faith in a provident god strengthens the immune system in the same way as any other placebo. This is because of the connections between the nervous system and our defences (Sagan and Margulis 2003: 317) and could explain many supposedly miraculous healings. Moreover, religions constituted an element of ethnic cohesion in ancient societies, in which national beliefs were predominant. Although it may appear to be unrelated, this observation has much in common with bacterial autopoiesis, that is with the capacity of even the simplest organisms to create a boundary, a membrane without which the awareness of singular/plural and I/we is impossible. Indeed, in clear disagreement with many philosophical schools, some biologists have defended the existence of this type of self-consciousness in microbial life (Sagan and Margulis 2003: 313-314), unlike the anthropocentric perspective that only recognises this trait in man or at the most in some of the so-called superior animals (Eccles 1992: 193).

The evolutive function here understands the ministry of Phoenician priests as producers of adaptive memetic mutations for their believers. This Darwinian perspective is innovative: indeed, although the "scientific" knowledge of the Canaanite clergy has been pointed out, the biolo-

gical reasons that link this knowledge to the colonial expansion have gone unnoticed. Let us first remember that Darwinism assumes that evolution takes place through the selection of random mutations. This assumption implies that a specie will raise its chances in future environmental conditions in direct relation to the degree of variation present within its population. The lack of genetic diversity and the homogeneity of behaviour could thus lead to a dead-end for survival. A given population would therefore be grateful in the long run if it were to possess a heterogeneous background of genes and behaviours. In many kinds of bacteria, evolution has often met this challenge by giving them the ability to transfer horizontally the recently acquired genetic mutations. Faced with a hostile environment (in contact with antibiotics for example), they receive a beneficial genetic contribution from part of their own population character-rised by its high production of changes. These subpopulations of "inventors" are known as *hypermutators*. Following this pattern, the horizontal transmission of memes can be studied by means of the same biological perspective as in the evolution of non-human living beings. These mechanisms of information transference, that move genetic and memetic codes both horizontally and vertically, are processes which for Darwinism, given their inability to create exact replicas, contribute to widespread variation.

The temples played an important role in the Phoenician colonisation. The gods guaranteed the economic agreements met therein (Bunnens 1979; Aubet 1994: 142). The written texts and archaeology show that the foundation of the sanctuaries preceded in many cases that of the colonies themselves (Aubet 1994: 141). This behaviour is not exclusive to the Canaanites; it is the case also of the Greeks. In Tartessos, Phoenician sanctuaries have been known of for a long time, but recently two new particularly important sites have been discovered: the sanctuary of Ba'al Tsaphon in ancient *Caura* (Coria del Río) and that of Astarte at the El Carambolo (Camas), both located in the province of Seville. Of particular interest is the astronomical orientation of these sacred sites.

The sanctuary of Ba'al at Coria del Río has revealed a clay altar in the shape of a bull skin whose longitudinal axis is directed towards the East to the sunrise of the summer solstice and to the West to the sunset of the winter solstice (fig. 3.1). This orientation, that obeys the pattern observed in many Iberian, Greek and Phoenician temples (Esteban 2002: 94), was deliberate given that its axis is somewhat deviated from the axis of the room within which it is located. The same orientation was adopted by the oldest of the five temples, although the four later phases modified the norm due to urbanistic and topographic requirements. However, the helioscopic orientation was maintained at least in sanctuary III dated to the 7th century BC. Similar cases to the altar of Coria have been documented in many other Protohistoric altars, for instance that of Oral (Abad and Sala 1993: 179). At El

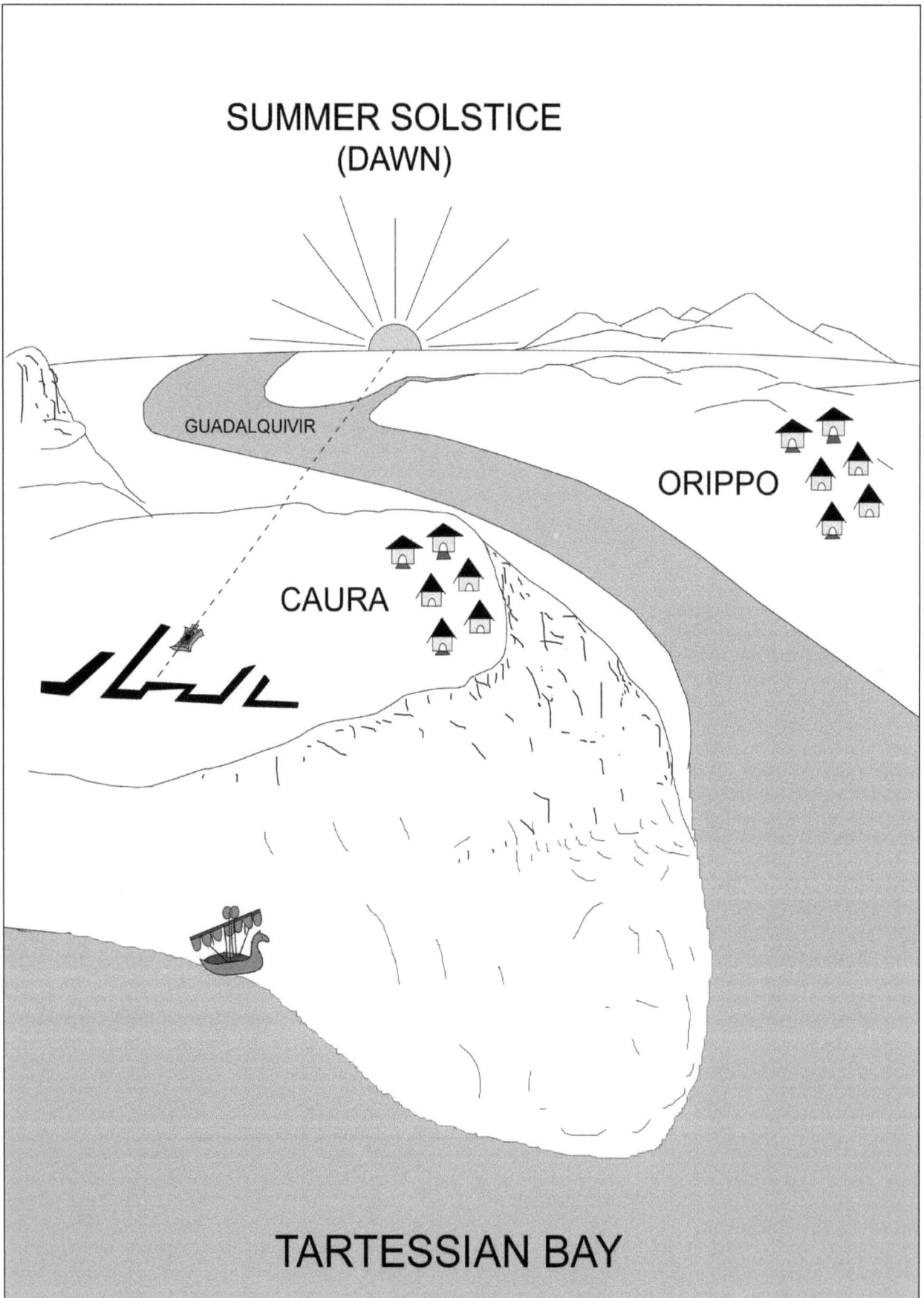

Fig. 3.1. The sanctuary of Ba'al at Coria del Río (Seville, Spain) has revealed a clay altar in the shape of a bull
skin whose longitudinal axis is directed towards the East to the sunrise of the summer solstice and
to the West to the sunset of the winter solstice

Carambolo, a luxurious building, occupying the entire hill top, has been discovered. Unlike the adjacent houses, it is also orientated towards the sun rise and set (its entrance faces East and its back entrance West). Although this complex originated in the 9th century BC with a more simple design, this solar orientation was present since its foundation and was respected in a later phase of enlargement (fig. 3.2). The *ex voto* of Astarte found at El Carambolo suggests that this temple may have been dedicated to this Phoenician goddess. However, the orientation of the building, with its entrance towards the sunrise of the summer solstice, suggests the greater importance of the male god for those who designed and ordered its construction, that is those who played a greater role in the cult: the priests. This could be a legacy of more ancient situations, since, despite the popular preference for Astarte-Anat in Ugarit in the late Bronze Age, the official Canaanite theology gave more importance to Ba'al (Liverani 1995: 452).

The first objective of the helioscopic disposition of the building may have been to fix the days in which to celebrate the festivities of the vital cycle of Ba'al. According to the later tradition that associated this god with Adonis, especially linked to a particular Ba'al of Byblos (Ribichini 2001: 105-106), the death and resurrection of the god were celebrated during the days of the summer solstice (Du Mesnil 1970: 108; Garbini 1965: 44) when the cereals were ripening and when the spring growth died back, struck like the god himself by the summer heat (Marlasca 2005: 458). The regulation of the calendar could thus be efficiently timed. The control of chronological time was in fact one of the abilities of Ba'al, assimilated to Cronus-Saturn from an early date (Bloch 1981: 127). The Phoenicians of Tartessos gave particular importance to this divine entity for which a temple was built in *Gadir*.

Fixing solstices in Antiquity was not without problems. For Ptolemaic science, the immobile nature of the sun implied a serious challenge for the fixing of these dates with precision. The documental history of astronomy maintains that the solution was reached in the Middle Ages, when the Islamic observatories carried out more precise measurements at other times of the year. However, archaeology shows that many prehistoric cultures were familiar with the solstitial phenomenon. In the case of the Phoenician altars of Tartessos, their immobile nature certainly helped the calculations. The greatest difficulty would then have been in determining the correct solstitial orientation during their construction.

The evolutive importance of this astronomical knowledge is related to the advance of the Phoenician colonial wave throughout the Mediterranean. In biological terms, the success or failure of individuals, of populations and of species can only be measured by their rate of reproduction and the alopatric expansion. This scale permits the classification of the mutations (genetic and memetic) as positive, negative or neutral according to whether they

contribute much, little or nothing to the demography. In the same way, a Darwinian perspective would recognise that a population with a wide scope of diversity would be better equipped to face future changes or unforeseen situations, and this if evolution was only an adaptative response to ecological succession. However, since evolutive processes are also characterised by genetic and behavioural modifications that can transform the environment to the advantage of the individual, population or specie that originated the transformation, the fact that some groups possess a subpopulation of hypermutators is an incomparable evolutive weapon. If the group has a mechanism that produces variation, the conditions of its own expansion become particularly ideal given the possibility that among the changes produced the ideal memes may coincide.

Consequently, the Phoenician clergy could have been, within its own society, one of the most dynamic sectors in the production of scientific memes. Thus among the symbolic, ritual and mythical complexity, that reminds us of the random creation of mutations within the genotype, astronomical knowledge that would be beneficial for the entire community emerged. The reason behind the evolutive benefits of such logical acquisitions explains why the sanctuaries were at the forefront of the expansion wave of the Phoenician colonisation.

During prehistory, navigation throughout the Mediterranean was generally limited to coast hopping. Out of sight of the coast, it was very difficult to establish return journeys; indeed, these contacts were more likely in the Aegean and other parts of the Eastern Mediterra-nean with frequent islands. For this reason, there is hardly any evidence of the Mycenaean to the West of Italy. Megaliths reveal knowledge of the cosmos since the Neolithic (Hoskin 2001) and it is likely that in the Copper Age some cultures were navigating with the stars and were able to carry out journeys through high sea. In any case, the collapse of the Chalcolithic world led to the loss in practise of this possible nautical tradition. In the 2nd Millennium BC, ships still guided themselves by the coastline. If they were lost, they used birds to locate land (Luzón and Coín 1986), a technique similar to that employed by Noah (*Gen.* 8: 6-11). The mapping of routes in the West initiated with the Greek periplus of the 6th century BC that was the inspiration for the *Ora Maritima* by *Rufius Festus Avienus*. However, the 1st Millennium BC brought about a drastic change: the Phoenicians introduced astronomically guided navigation (Pliny, *Nat. Hist.* VII, 209; Strabo, *Geog.* I, 1, 6). The new system made it easier to plan journeys by sea, thus motivating intercommunity contacts and the subsequent increase in diversity in many regions. Evolutive theory knows well that the rate of change increases in proportion to variation, since greater variability provides natural selection with more alternatives (Ayala 1994: 67). Any historian familiar with evolutionism will recognise this as the reason behind the drastic and rapid changes in various Mediterranean cultural contexts that took place in the 1st Millennium BC.

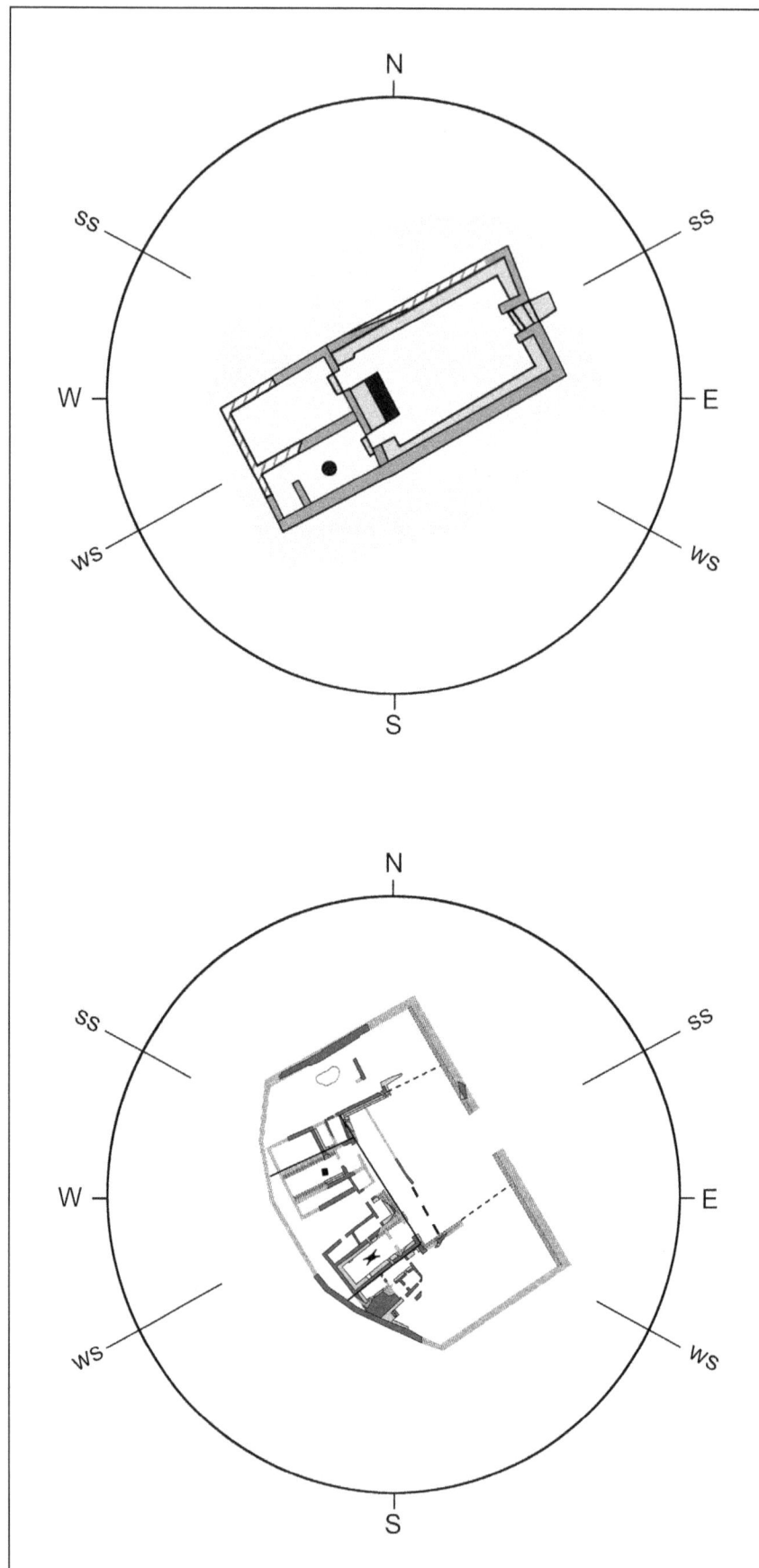

Fig. 3.2. The Carambolo compound is also orientated towards the sun rise and set. Although it was originated in the 9th century BC with a more simple design, this solar orientation was present since its foundation and was respected in a later phase of enlargement

The use of new nautical procedures was made possible by the existence of observations which, under the theological appearance of the knowledge of the divine beings (Ba'al was assimilated to the solar disc as a god and an omnipotent star, Astarte was identified with planet Venus), the Phoenician clergy had acquired in the temples. For this reason, among others, the populational expansion required the creation of sanctuaries in the main colonial centres. For a similar reason, many of these cult centres were built in coastal locations, places that helped the fluid transmission of information between seamen and priests. Moreover, the number of sanctuaries located along the coast displays their utility and explains why many of these sacred places were not located with the urban areas. This interpretation also explains why the colonial foundations enabled by maritime expeditions were accompanied by sacred oracles.

The rhythm and quantity of memetic mutations of this kind are directly proportional to the effort invested by the community, measured as the number of people and the time employed. Today, this ratio could easily be calculated since the budgets dedicated to research by states and other institutions are well documented. However, regarding the idea that evolutive theory can only explain the pleistocenic past and, at best, only in terms of corporal modifications, few modern historians have taken into account the natural processes involved in this matter, that translate in terms of long term reproductive benefits for the group. In this sense, if it is costly in the short term to exempt part of a population from the direct production of material goods, evolution would have tended to develop means of avoiding the failure of such a decision. As we shall show, this condition was met through the adoption of boundaries that restricted the transmission of new knowledge to other communities that had no made the effort of investment.

T. Chapa and A. Madrigal (1997: 189-190) have reviewed some of the main exemptions of priests in several cultures of the ancient world. The common denominator was the exemption from military obligations. However, priests were also dispensed from agricultural tasks, work on board ships and many other tasks involving manual labour. To maintain this subpopulation merely as memetic hypermutators would have been expensive if their achievements had not have been worthwhile. The parallel emergence of means of encrypting the memetic mutations is thus predictable, with the purpose of stopping the adaptative results of some of the mutations from crossing the group boundaries into other populations.

A generic boundary was the spontaneous tendency towards the ramification of religions. By creating "heathen", those who do not follow the same beliefs, each religion shapes, intentionally or not, a series of intercommunitary boundaries. It is possible that this splitting tendency is but a manifestation of the 2nd law of thermodynamics, which recognises that throughout the cosmos, entropy is constantly increasing and is a principle

that also affects life on our planet (Atkins 1992: 33). In any case, the memes selected by priests of religions that consider each other as mutually "pagan", and false with respect to the authentic credo, possess little possibility of penetrating in those who do not profess the same faith. The biblical contempt towards the beliefs of the Canaanites is an example of such a boundary. The fundamental hindrance to address this question is the sparse archaeological record that they leave behind. But something can be done.

Today, patenting is the mechanism that establishes the ownership of a scientific or technical discovery. The bond between invention and patent is so strong that one cannot be imagined without the other. This protection guarantees that the author or sponsor of the discovery will benefit from the corresponding earnings.[1] In the case of the astronomical knowledge of the Phoenician clergy, the possible leakage from this community was avoided in great part thanks to writing, essentially because among its uses, the ceremonial was limited to specialists of the cult[2] (Oppenheim 2003: 222). At first sight, this affirmation may seem paradoxical since graphical systems are used to communicate. However, given that their use was very restricted in ancient cultures, putting a message into writing ensured its limited circulation. This observation has been made by specialists in various prehistoric writing forms in *Hispania* who underline their "somewhat esoteric" character (De Hoz 1989: 549). Because of their capacity to close the spectrum to which new knowledge is passed, it is not surprising that several oriental graphical systems emerged in temples, although their use was not restricted to this function. In Tartessos, perhaps the oldest example of writing precisely comes from a sanctuary. It is the text inscribed at the feet of Astarte of the Carambolo, in which the devotees give thanks for a granted favour. The text contains no practical knowledge of astronomy nor anything that may be considered as scientific, but its presence in a sacred site suggests that it may have been written and understood within this sphere. We shall now turn to the restrictive geography of the use of writing, that

---

[1] Some ancient traditions my have enclosed mechanisms of this kind. For example, in the case of Greek mythology and specifically the myth of Hephaestus, a possible interpretation of the tradition according to which the "Smith-god hobbles" -that spreads through very different regions such as Western Africa or Scandinavia- suggests that in ancient times the smiths were crippled intentionally to avoid them going to enemy tribes and passing on their knowledge (Graves 1981: 88). The nine year stay of Hephaestus in the cave of Lemnos could be related to with this secret objective, once metallurgy arrived to Greece from the Aegean Islands.

[2] It is worth recalling the passage prior to the initiation of Lucius in the *Metamorphoses* when, accompanied by the priest in charge of the cult of Isis, he enters one of the most secret rooms of the sanctuary: «Then that kindest of men took me by the hand and led me straight up to the entrance of the great temple. After the ceremony of opening had been celebrated with the prescribed ritual and the morning sacrifice had been completed, he brought out from the secret part of the sanctuary some books inscribed with unknown characters. Some used the shapes of all sorts of animals to represent abridged expressions of liturgical language; in other, the ends of the letters were knotted and curved like wheels or interwoven like vine-tendrils to protect their meaning from the curiosity of the uninitiated» (Apuleius, *Metamorphoses*, XI, 22).

is turn leads us to agree with those who argue the illiterate nature of the majority of the Andalusian prehistoric population (Chic 1999: 179).

If writing may, willingly or not, hide memetic mutations, its diversity itself may be read from an evolutive perspective as a further insistence in this direction. Thus the political mosaic characteristic of the oriental city state systems, which was replicated in the Phoenician colonial territories, constituted the ideal ecosystem for this evolutive radiation, limited only by the need of a common language and graphical system for the commercial relationships. The resource that avoided in the ancient Near East the unlimited circulation of scientific advances was the use by the priests of a language that the rest of the community did not understand, in many cases an ancestral form of the everyday language used by the population. This practice is known from an endless list of Asian cases and from the Egyptian world. However, given the lack of evidence, it is impossible to confirm if that this was the case of the Phoenicians of the 1st Millennium BC. Without confusing writing with language, the graphical system known as Tartesian, probably originated in religious spheres, coping some signs of the ancient Phoenician alphabet, more archaic that those used by the Canaanite colons when they arrived to this territory in the 9th century BC. Perhaps the contradictions pointed out by J. de Hoz (1986: 76, 80-82) between the earliest dates of the colonisation of *Hispania* and the dates of the expansion of graphical systems can find an answer in this explanation of the evolutive role played by the Phoenician priests, according to which they would have used a liturgical writing different from that commonly employed at the time. However, it is impossible to say whether they also used a different language. A similar cryptographic procedure was used in Egypt of the pharaohs (Hornung 1992: 33-34) and is used by the present-day Coptic Christian priests in Ethiopia.

It seems obvious that if the Phoenician implantation in Iberia was so deep, this was enabled directly by the extraordinary development of their maritime routes. In this expansion, the temples were particularly important because therein was developed the astronomical knowledge necessary for navigation. If the science of the sky enabled the Phoenicians to acquire important knowledge about the position and movement of the stars and the organisation of the calendar, in parallel it served the exclusive expansion of their own group. As predicted by the Darwinian perspective, there could have existed barriers that hindered the access of "heathen" (members of other communities) to these sacred places since these were where the transmission of knowledge and practical applications took place between the priests and the seafarers. The custom that only admits the entrance to the temples to those that profess the same religion is still today practiced in some confessions. The most plausible evolutive explanation for this lies in the limits established by natural selection for the horizontal memetic transmission of positive mutations, something that is

well-known in the case of the interindividual genetic transference of bacteria and plasma. This limitation does not however necessarily signify that non-Phoenician people could not enter the sanctuaries where the astronomical experience was stored and from which the oracles for the foundation of new colonies were emitted, that is those that were during some time at the forefront of the demographic expansion. This could have been the case in some of the more humble temples but is less likely in the case of those where the knowledge was safely kept. It is therefore quite unlikely that the indigenous people of Tartessos could have accessed the Carambolo for example, the most important Phoenician sanctuary excavated in the West until the date. However, when the sea routes were common to Phoenicians and Greeks, in the second half of the first millennium B.C., the role that evolution had reserved for the astronomical knowledge of the priests went out of use, partly because there were not many new areas to explore: neither virgin territories nor accessible routes. In the Hellenistic period, the temple of Melkart in Cadiz was opened to Greek scholars (Marín and Jiménez 2004: 227-228). This explains that the *mqm'lm* ("resuscitator or the divinity"), the priestly duty that accumulated most astronomical expertise (Escacena 2006: 146), lost the importance that it had had during the phase of colonial expansion. However, since one of the singular characteristics of the clergy since its origin was to be the motor of variation, especially obvious in the diversity of its functions, it was precisely this heterogeneity of ecological niches that insured its later existence, evermore linked to what has been named "ethical religion" in the place of "cultual religion" (Alonso 2003: 460-462).

CONCLUSIONS

The adaptative mechanisms that preserve the positive memes for the exclusive use of the group emerge for strong evolutive reasons and revalidate the well known Darwinian experience by which natural selection rarely acts upon the entire species but upon parts of it. These fractions, known in biology as *populations*, correspond to terms used by historians and archaeologists such as *ethnic group, nations, people or human group*. The comparison between the Phoenician clergy and bacterial subpopulations that act to their own benefit as hypermutators, equipped with random genetic changes, some of which could become adaptative in new contexts and manage to maintain the growth of the population, suggests that the horizontal transference of genes and memes does have boundaries. Also in relation with human behaviour, natural selection, in this case as a mechanism of group or interdemic selection, has created inhibitive filters to the free circulation of memes from the inventor populations towards other distinct groups.

The most frequent posture among specialists is to consider the priests of ancient oriental cultures as a key piece in the maintenance and reproduction of social inequality (Liverani 1995: 119). Moreover, for Evolutive Archaeology, the assessment of the historical importance of the ancient clergy must be carried out from the standpoint of its contribution to demographic growth and to the subsequent dispersion of the communities of which these specialists were part. These two variables (demographic growth and geographic expansion) represent ideal indicators of the fitness of individuals and populations.[3] From this perspective, the Phoenician clergy played an important role in the diaspora of its community: it was the holder and guarantor of the astronomical knowledge necessary for long haul nautical navigation, as well as the researchers in this scientific enterprise. Not in vain, the foundation of many important colonies was accompanied, when not preceded, by the appropriate consecration of sanctuaries. It is not at all meaningless from an evolutive point of view that there are known cases of colonising expeditions preceded by their respective oracles, such as the oracle of Posidonius recorded by Strabo regarding the foundation of Cadiz.

Acknowledgements

We would like to thank our colleague Ruth Taylor for her generous and essential help with the English translation.

Bibliography

ABAD, L.; SALA, F. (1993): *El poblado ibérico de El Oral (San Fulgencio, Alicante)* (Trabajos Varios del S.I.P. 90). Valencia: Diputación de Valencia.

ALEXANDER, R. (1994): *Darwinismo y asuntos humanos*. Barcelona: Salvat. (1979: *Darwinism and human affairs*. Seattle: University of Washington Press).

ALONSO, J. (2003): "Religión cúltica y religión ética (en tensión desde la Biblia)", in A. González, J.P. Vita and J.A. Zamora (Eds.): *De la tablilla a la inteligencia artificial*, vol. II: 457-476. Zaragoza: Instituto de Estudios Islámicos y del Oriente Próximo.

ALVAR, J. (1993): "Problemas metodológicos sobre el préstamo religioso", in J. Alvar, C. Wagner and C. Blánquez (Eds.): *Formas de difusión de las religiones antiguas. Segundo encuentro-coloquio de ARYS*: 1-33. Madrid.

---

[3] Talking about the clergy, in the specific case of Christianity, it is interesting that two basic constants throughout the Old Testament are precisely the promise of multiplication and widespread diffusion of the community chosen by their God. Regarding the religious dimension itself, the hope of salvation always gives survival to a numerous group. For example, we could mention the parable of the sacrifice of Abraham. Once the angel of Yahveh aborts the death of Isaac, he promises Abraham the multiplication of his descendents and their diffusion and conquest of new lands: "... That in blessing I will bless thee, and in multiplying I will multiply thy seed as the stars of the heaven, and as the sand which is upon the seashore; and thy seed shall possess the gate of his enemies; and in thy seed shall all the nations of the earth be blessed..." (Genesis 22, 15-18).

APULEIUS, *Metamorphoses*, vol. II. Edited and Translated by J. Arthur Hanson. 1989. Cambridge: Harvard University Press.

ATKINS, P.W. (1992): *La segunda ley*. Barcelona: Prensa Científica. (1984: *The Second Law*. New York: Scientific American Books).

AUBET, M.E. (1994): *Tiro y las colonias fenicias de Occidente*. Barcelona: Crítica.

AYALA, F.J. (1994): *La teoría de la evolución. De Darwin a los últimos avances de la genética*. Madrid: Temas de Hoy.

BLACKMORE, S. (2000): *La máquina de los memes*. Barcelona: Paidós. (1999: *The meme machine*. Oxford-New York: Oxford University Press).

BLOCH, R. (1981): "Le culte étrusco-punique de Pyrgi vers 500 avant J.C.", *Die Göttin von Pyrgi. Archäologische, linguistische und religionsgeschichtliche Aspekte*: 123-129. Firenze: L.S. Olschki.

BOYD, R.; SILK, J.B. (2001): *Cómo evolucionaron los humanos*. Ariel, Barcelona. (2000: *How humans envolved*. New York: Norton).

BUNNENS, G. (1979): *L'expansion phénicienne en Méditerranée*. Bruxelles-Rome: Institut Historique Belge de Rome.

BURKERT, W. (1996): *Creation of the sacred. Tracks of biology in early religions*. Cambridge: Harvard University Press.

CHAPA, T.; MADRIGAL, A. (1997): "El sacerdocio en época ibérica", *Spal* 6: 187-203.

CHIC, G. (1999): "Comunidades indígenas en el Sur de la Península Ibérica: dos notas", in F. Villar and F. Beltrán (Eds.): *Pueblos, lenguas y escrituras en la Hispania prerromana*: 173-182. Zaragoza-Salamanca: Institución "Fernando el Católico" – Universidad de Salamanca.

DE HOZ, J. (1986): "Escritura fenicia y escrituras hispánicas. Algunos aspectos de su relación", in G. del Olmo and M.E. Aubet (Dirs.): *Los fenicios en la Península Ibérica*, vol. 2: 73-84. Sabadell: Ausa.

DE HOZ, J. (1989): "El desarrollo de la escritura y las lenguas de la zona meridional", in M.E. Aubet (Coord.): *Tartessos. Arqueología protohistórica del Bajo Guadalquivir*: 523-587. Sabadell: Ausa.

DENNETT, D.C. (1998): "The evolution of religious memes: who or what-benefts?", *Method and Theory in the Study of Religion* 10: 115-128.

DIAMOND, J. (2001): "La evolución de los gérmenes y las armas de fuego", in A.C. Fabian (Ed.): *Evolución. Sociedad, ciencia y universo*: 77-102. Barcelona: Tusquets.

DU MESNIL, R. (1970): *Études sur les dieux phéniciens hérités par l'empire romain*. Leiden: E.J. Brill.

ECCLES, J.C. (1992): *La evolución del cerebro: creación de la conciencia*. Barcelona: Labor. (1989: *Evolution of the Brain: Creation of the Self*. London: Routledge).

ESCACENA, J.L. (2006): "Allas el estrellero, o Darwin en las sacristías", in J.L. Escacena and E. Ferrer (Eds.), *Entre Dios y los hombres: el sacerdocio en la Antigüedad* (Spal Monografías VII): 103-156. Sevilla: Universidad de Sevilla.

ESTEBAN, C. (2002): "Elementos astronómicos en el mundo religioso y funerario ibérico", *Trabajos de Prehistoria* 59 (2): 81-100.

GARBINI, G. (1965): "Considerazioni sull'inscrizione punica di Pyrgi", *Oriens Antiquus* 4: 35-52.

GOULD, S.J., VRBA, E.S. (1982): "Exaptation – a missing trem in the science of form", *Paleobiology* 8: 4-15.

GRAVES, R. (1981): *The Greek Myths*, vol. 1. Harmondsworth: Penguin Books.

HART, J.P.; TERRELL, J.E. (Eds.) (2002): *Darwin and archaeology: a handbook of key concepts*. Bergin & Garvey, London.

HORNUNG, E. (1992): *Idea into image. Essays on ancient Egyptian thought*. New York: Timken.

HOSKIN, M. (2001): *Tombs, temples and their orientations. A new perspective on mediterranean Prehistory*. Bognor Regis: Ocarina Books.

LINCOLN, B. (1981): *Priests, warriors & cattle: Study in the Ecology of religions*. University of California Press.

LIVERANI, M. (1995): *El antiguo Oriente. Historia, sociedad y economía*. Barcelona: Crítica. (1988: *Antico Oriente: storia, società, economia*. Bari Laterza).

LUZÓN, J.M.; COÍN, L. (1986): "La navegación pre-astronómica en la Antigüedad: utilización de pájaros en la orientación náutica", *Lvcentvm* V: 65-85.

MASCHNER, H.D.G. (Ed.) (1996): *Darwinian archaeologies*. New York: Plenum Press.

MARGULIS, L. (2003): *Una revolución en la evolución*. Valencia: Universidad de Valencia.

MARÍN, M.C.; JIMÉNEZ, A.M. (2004): "Los santuarios fenicio-púnicos como centros de sabiduría: el templo de Melqart en Gadir", *Huelva Arqueológica* 20 (Actas del III Congreso Español de Antiguo Oriente Próximo): 217-239.

MARLASCA, R. (2005): "La *egersis* de Melqart: una propuesta para su interpretación", en A. Spanò (ed.), *V Congresso Internazionale di Studi Fenici e Punici* (I): 455-461. Palermo: Università degli Studi di Palermo.

OPPENHEIM, A.L. (2003): *La antigua Mesopotamia: retrato de una civilización extinguida*. Gredos, Madrid. (1970: *Ancient Mesopotamia: portrait of a dead civilisation*. Chicago: University Press).

PUNSET. E. (2004): *Adaptarse a la marea. La selección natural en los negocios*. Madrid: Espasa Calpe.

QUEROL, M.A. (2001): *Adán y Darwin*. Madrid: Síntesis.

RIBICHINI, S. (2001): "La scomparsa di Adonis", in P. Xella (Ed.), *Quando un dio muore. Morti e assenze divine nelle antiche tradizioni mediterranee*: 97-114. Verona: Essedue.

RINDOS, D. (1984): *The Origins of Agriculture: an Evolutionary Perspective*. New York: Academic Press.

RINDOS, D. (1988): "Evolución darviniana y cambio cultural. El caso de la agricultura", in L. Manzanilla (Ed.), *Coloquio V. Gordon Childe. Estudios sobre la Revolución Neolítica y la Revolución Urbana*: 79-90. México: Universidad Nacional Autónoma de México.

SAGAN, D.; MARGULIS, L. (2003): "El yo ininterrumpido", in L. Margulis, *Una revolución en la evolución*: 301-318. Valencia: Universidad de Valencia.

TOBEÑA, A. (2005): *Mártires mortíferos. Biología del altruismo letal. Un itinerario por el cerebro de los suicidas atacantes*. Valencia: Universidad de Valencia.

WYNNE-EDWARDS, V.C. (1963): "Intergroup selection in the evolution of social systems", *Nature* 200: 623-626.

AN EVOLUTIONARY THEORY OF CULTURAL DIFFERENTIATION

Agner FOG

Aalborg Univ. Copenhagen / Copenhagen Univ. Coll. of Engineering; Lautrupvang 15, 2750 Ballerup, Denmark;
anger@agner.org , www.agner.org

Abstract: *This paper introduces the cultural r/k theory – an evolutionary theory that explains why different cultures evolve in different directions. The cultural r/k theory links differences in artistic style with war and peace, geography, political system and religion. This theory is useful for explaining cultural differences, for classifying artefacts and for predicting sampling bias in the archaeological record.*
Keywords: *cultural r/k theory- evolutionary theory*

Résumé: *Ce travail propose d'introduire la théorie culturelle r/k – une théorie évolutive qui explique pourquoi les différentes cultures se développent dans des directions différentes. La théorie culturelle r/k relie les différences stylistiques de l'art avec la guerre et la paix, la géographie, le système et la religion. Cette théorie est utile pour expliquer les différences culturelles, pour classifier les objets et pour prédire les biais produits dans l'échantillonnage du registre archéologique.*
Mots clés: *théorie culturelle r/k – théorie évolutive*

The sciences of palaeontology and evolutionary biology have contributed very much to each other. A similar link between archaeology and theories of cultural evolution is much weaker. This lack of synergy is probably due to the poor state of cultural evolution theory. The various theories of cultural evolution have been criticized for more than half a century for being unilinear, teleological and ethnocentric. The theories hardly make the distinction between evolution and progress, and it is assumed that cultural evolution can only go in one direction.

The problem with cultural evolutionism is, in my opinion, that it has focused too much on defining the direction of evolution and too little on theoretical models of mechanisms that can explain why evolution goes in a particular direction.

I will therefore present a new theory, which can explain why different cultures evolve in different directions. This theory is called *cultural r/k theory* after a weak analogy with the so-called r/K theory in evolutionary biology (Fog, 1999). Cultural r/k theory defines two opposite directions of evolution called *regal* and *kalyptic*. The regal direction is characterized by bellicosity, strict discipline, intolerance and authoritarianism. The kalyptic direction is characterized by peacefulness, tolerance and individualism. A culture is driven in the regal direction by war, threats of war, or any other danger that is perceived to threaten the cultural group as a whole. A culture will evolve in the kalyptic direction in the absence of any such collective threats. The regal/kalyptic continuum is just one of a number of dimensions that can be used for characterizing a culture. Other useful dimensions are: degree of technological sophistication, political organization, and specific adaptation versus general adaptivity. The present paper will focus on the r/k dimension and its possible applications in archaeology.

The placement of a culture on the r/k scale is reflected in its political system, religion, philosophy, morals and justice system. What is particularly interesting to archaeology is that the r/k dimension is also reflected very strongly in the artistic production of a culture, including art, architecture and music.

The characteristic signs of regal and kalyptic tendencies in different spheres of life are briefly summarized in the table below.

The most regal cultures are typically found in areas where the geography and climate facilitate war and a high population density. This includes large continents with fertile ground and a moderate climate, where easy means of transportation exist. Examples of regal cultures in historic time are the great empires in Europe, Arabia, Asia and precolonial America.

The most kalyptic cultures are found on isolated islands where there are no enemies, in areas where the natural resources cannot sustain a high population density, in areas where dense vegetation and absence of efficient transportation means limit the possibilities of warfare, and in very hot or very cold climates where warfare is difficult. Examples are found on Pacific islands, in Arctic areas and sub-Saharan Africa.

The rise and fall of big empires can be explained in terms of cultural r/k theory. The growth of a city-state into a state and then an empire is a process of regalization. The chief or ruler may decide to increase the military strength of the group for defensive or offensive reasons. Improvements in military technology, such as horses or gunpowder may contribute to the process. Improvements in food production technology, such as the introduction and spread of agriculture leads to increased population density and the possibility for increasing the size of the army. Population growth creates a need for enlarging the territory. The growing military strength of the group makes it possible to defeat the more kalyptic neighbour states and incorporate them into the empire. The regal

Tab. 4.1. Regal and kalyptic characteristics in various spheres of life

Sphere	Regal	Kalyptic
Philosophy	Individuals exist for the benefit of society. Ethnocentrism, racism, growth, expansion.	Society exists for the benefit of the individual. Individualism, tolerance, human rights, protection of natural resources.
Religion	Monotheism, polytheism. Fundamentalist, ascetic, puritan.	Animism, pantheism, atheism, fertility cult, ancestor worship.
Politics	Powerful central government. Imperialism, uniformity, intolerance, slavery, censorship, severe punishments.	Decentralized or non-hierarchical government. Democracy, tolerance, peace.
Sexual behaviour	Strict sexual morals, stereotypical sex roles. Sex is only for procreation. Procreation is a duty. Early marriage. Polygamy. Contraception and abortion illegal. High population growth.	Liberal sexual morals. Sex is for pleasure. Flexible, individual behaviour. Contraception and abortion accepted. No population growth.
Education	Childhood is short. Little or no education. Children begin to work at an early age.	Long childhood. Parenting and education is important.
Art	Finical, perfectionist, very richly embellished. Repetition of small details with strict geometry. Portrays symbols of power such as gods, rulers, war heroes or predators.	Unrestrained, improvised. Depicts pleasure, fantasy, colours, animals, fertility, individualism, rebelliousness.
Architecture	Religious and government buildings are grandiose, ostentatious, rich in details, with oversized gates and towers.	Functionalistic, creative, individualistic, irregular. No stylistic demonstration of social status differences.
Dress	Decent, tidy, uniform. Differentiated after sex, social status, and group identity.	Creative, individual, colourful, sexy. Reflects personal taste.
Music and singing	Small variations in pitch. Rich and fine embellishment. By offensive regality pompous. Strict rules for rhymes and foot. Choir singing. Litany. Praises gods, rulers, military superiority or true love.	Bass accompaniment dominates over melodic voice. Rhythmic, varied, imaginative, often improvised. Broad repertoire of text themes.
Dance	Organised, restrained.	Unorganised, hilarious.

culture is imposed upon the annexed populations, whereby the empire grows. This process continues until the empire has reached the maximum size that is practically manageable. The size of the empire is limited by the available communication technology and other practical factors. It is difficult to lead a war on a distant boundary of the territory where communication with the ruler is slow and unreliable. And it is difficult to defend all borders on the large circumference of the territory at the same time.

The process of kalyptization starts when the empire has reached the maximum manageable size. It is difficult to motivate people to sacrifice big resources on a war that takes place so far away that it seems irrelevant. Only the most despotic government is able to keep together such a huge empire and maintain the necessary discipline and military strength. The population cannot see the necessity of a highly tyrannical rule, so they start to disobey and rebel. When the emperor is overthrown or reluctantly begins to loosen his iron hand, then the internal conflicts start to flare up. All those subgroups which, one by one, had been incorporated into the empire, have preserved some of their religious or ethnic identity. This identity is reinforced by their urge for independence and their

rebellion against the despotic rule. The allegiance to the ruler is gone and the empire starts to disintegrate into smaller groups. Some of these smaller groups may be incorporated into another growing empire.

The r/k processes can be summarized in the simple principle that regalization is characterized by inter-group conflicts, while kalyptization is characterized by intra-group conflicts.

The theoretical background of the r/k theory can be explained in terms of evolutionary psychology and the theory of vicarious selection. Let me first explain what vicarious selection is. The best example of vicarious selection is cultural evolution. The capacity for culture has been created by biological evolution. The high capacity for cultural adaptation gives the human race a much better adaptivity than other species because cultural evolution goes much faster than biological evolution. As long as the cultural evolution goes in approximately the same direction as biological evolution, i.e. towards increased capacity for survival, then it makes sense to consider the cultural evolution as vicarious for biological evolution. Biological evolution has created its own substitute, which enables it to reach its goal more efficiently.

The cultural r/k selection can also be considered a vicarious adaptation process. Consider the situation where a peaceful tribe with a flat political structure is surrounded by belligerent neighbour tribes with a hierarchical rule and strict discipline. If the peaceful tribe does nothing to prepare themselves for meeting the threat then they will soon be overpowered by the stronger neighbours, who will either annihilate them or impose their political system on them. The only way to avoid a cruel fate is to produce more children, strengthen the discipline and arm politically as well as morally. Whatever they do, the evolutionary result will be the same: that the political system that leads to the strongest military power will spread in the region. The political and moral armament that we call regalization will reach the same result faster and with fewer costs in terms of deaths than war. The psychological selection process that we call regalization is therefore vicarious for the more costly selection process called war.

The opposite process also needs an explanation. A tribe with a strict discipline and a despotic rule is not the best adaptation to a peaceful environment. The strict discipline takes a high toll on all members of the group. The unification of thought prevents the individual initiative and inventiveness that is the root of adaptation. And perhaps most importantly, the unrestrained population growth and greed may lead to an exhaustion of the natural resources that the group relies on for its subsistence. The social psychological mechanism that we call kalyptization can lower the costs to the individual and prevent exhaustion of the natural resources. This selection process is vicarious for the more costly selection process that takes place if the group perishes for failure to adapt to and preserve its environment.

My claim is now that the psychological process that I have called cultural r/k selection is an adaptation mechanism that has evolved by biological evolution as a vicarious selection process. This psychological mechanism allows a group to adapt to a changing environment faster and with fewer costs than by allowing the unfit cultures to perish. It is possible to envision many different scenarios that can account for the advantage of such a vicarious mechanism. The scenarios that I have sketched above are just the ones that I find most appropriate for what we know about human prehistory.

The cultural r/k theory is not the only available theory of why cultures change in politically strict or lax directions. The theory of the authoritarian personality makes the same connection between collective dangers and political strictness (Adorno, et. al., 1950). However, the theory of authoritarianism has been criticized for being politically biased because it attaches a label of psychopathology to certain political movements (McKinney, 1973). I am proposing the cultural r/k theory as a much-needed replacement for authoritarianism theory because it has a more sound theoretical basis and because it fits into the diverse paradigms of evolutionary psychology, cultural

evolution, social psychology, political history and culture studies.

Having explained the theoretical basis of this theory, I will now return to the factors that have driven cultures in regal or kalyptic directions throughout history. The most important factor pushing in the regal direction is, of course, war or threats of war. But other collective dangers such as natural disasters, famine and economic crisis have, at different times in history, been strong contributors to regal developments. It makes no difference to the psychological effect whether the collective danger is real or imagined. Political and religious leaders have often attempted to strengthen their dwindling power by exaggerating the seriousness of various dangers or by inventing fictive dangers. This is known as witch-hunting.

Kalyptic developments have been prevalent in periods of peace. This process is self-amplifying because the rate of population growth is lowered in kalyptic times whereby the impetus for war is removed. Economic interdependence between groups has also been a stabilizing factor preventing armed conflict.

The cultural r/k theory is no less applicable to modern societies. The increased economic interdependence between countries around the World as well as international peace-keeping efforts are strong factors leading our society in the kalyptic direction. These factors are, however, balanced by almost equally strong regalizing factors. Mass immigration from foreign cultures creates a fear of strangers in many countries in the world. Various witch-hunts and periodic economic crises are also regalizing factors. The mass media have a penchant for focusing more on bad news than good news. Crime, disaster, war and terror – wherever in the World it may happen – are presented in the news every day in such amounts that TV viewers and newspaper readers come to perceive the World as more dangerous than it is. This is perhaps the strongest regalizing factor in modern society today.

Let me finish this introduction to cultural r/k theory by discussing its application to archaeology. The r/k theory provides an easy means for classifying cultural artefacts. Artefacts with rich and perfectionist embellishment and strict geometry are signs of regal cultures, while artefacts depicting unrestrained fantasy and variation are signs of kalyptic cultures. These signs can be compared with what is known about the geography, climate and the political organization of the culture that produced the artefact.

The cultural r/k theory also reveals that archaeology can be expected to have a strong *sampling bias*. Regal cultures produce big and impressive artefacts made of durable materials, while kalyptic cultures produce smaller artefacts made of perishable materials. The most impressive remnants of past regal cultures are very likely to be found, while the remnants of the most kalyptic cultures of the past may have decomposed long ago.

Whoever finds a half-decomposed and not very impressive artefact may not appreciate its scientific value and preserve it. The bias is worsened by the fact that regal cultures are likely to systematically destroy artefacts left over from previous kalyptic cultures (This has happened even in modern times in Nazi Germany and Afghanistan). Kalyptic cultures, on the other hand, are likely to preserve and possibly even admire and collect the artefacts left from previous regal cultures. Regal artefacts can thus be found in kalyptic cultures, but not vice versa. This sampling bias cannot be offset by the fact that regal artefacts are more tempting objects of looting.

More details about the cultural r/k theory can be found in Fog (1999).

References

ADORNO, T.W.; FRENKEL-BRUNSWIK, E.; LEVIN-SON, D.J.; SANFORD, R.N. (1950) – *The Authoritarian Personality*. New York: Harper & Brothers.

FOG, A. (1999) – *Cultural Selection*. Dordrecht: Kluwer.

MCKINNEY, D.W. Jr. (1973) – *The Authoritarian Personality Studies*. Haag: Mouton.

A GROUP SELECTION MODEL OF TERRITORIAL WAR, XENOPHOBIA AND ALTRUISM IN HUMANS AND OTHER PRIMATES

Agner FOG

Aalborg Univ. Copenhagen / Copenhagen Univ. Coll. of Engineering, Lautrupvang 15, 2750 Ballerup, Denmark, anger@agner.org, www.agner.org

Abstract: A theoretical model of wars over group territories shows that behavioural traits like cooperative warfare, justice, altruism and outsider exclusion may have coevolved in higher primates and prehistoric man. The conditions for territorial war to be an effective mechanism of group selection are discussed. These conditions may have been present in tribal societies in prehistoric times but not in modern times. The geographic evolution of territories is illustrated with computer simulations.
Keywords: group selection- model

Résumé: Un modèle théorique des guerres à travers les territoires des groupes montre que les traits de comportements tels que la guerre coopérative, la justice, l'altruisme et l'exclusion de l'étranger peuvent avoir évolués ensemble chez les grands primates et chez l'homme préhistorique. Les conditions pour la guerre territoriale comme mécanisme efficace de la sélection de groupe sont discutées. Ces conditions ont du être présentes chez les sociétés tribales des temps préhistoriques mais pas des temps modernes. L'évolution géographique des territoires est illustrée par des simulations informatiques.
Mots clés: sélection de groupe – modèle

Altruism means unselfishness. Altruism can be defined in evolutionary theory as a behavioural trait that increases the genetic fitness of others but decreases the fitness of the actor, who is called altruist. This is a very controversial issue in evolutionary biology because the principle of "the survival of the fittest" would predict that altruism is eliminated by natural selection.

Altruistic behaviour is nevertheless widespread in human societies. The almost universal readiness to help strangers in need and the widespread support for charity organizations is evidence that altruism is indeed a human trait. A number of possible theoretical explanations have been proposed. The most important explanations are the following theories:

- Group selection. A group of altruists is more likely to survive in hard times than a group of egoists. The egoists may be eliminated by extinction of entire groups (tribes) or even species.

- Reciprocal selection. It may be profitable for one individual to help another if it can be expected that the favour is later returned. This mechanism has been extensively studied in game theory, where it is known as the prisoner's dilemma. The theory involves many complications such as the probability of meeting again, cheating, and the possibilities of detecting cheating.

- Kin selection. Natural selection will promote any behaviour that leads to the production of more copies of the gene that codes for this behaviour. This includes the helping of others who share the same gene. A gene for helping one's siblings will spread if the fitness gain to the sibling is more than double the cost to the helper, because full siblings have a 50% probability of sharing the same gene. Helping distant relatives is less likely to be promoted by kin selection because they have a low probability of sharing the same gene.

- Cultural reward and punishment. Altruists are rewarded with a good reputation, which may confer various social advantages that increase their genetic fitness. Egoists may be punished as criminals.

- Cultural selection. Cultural and religious norms that command their adherers to help others are likely to spread by cultural selection. It is easy to observe that charity is often connected with religion, but there is no evidence that atheists are less altruistic than devoutly religious persons.

All of these theories probably contain part of the explanation why altruism is widespread among humans. Group selection is the theory that can explain the most examples of altruistic behaviour, but also the most controversial theory.

Various mathematical models have shown that the individual selection for egoism will be stronger than the group selection for altruism in almost all cases (Boorman and Levitt, 1980). However, this theory is in sharp contrast to actual observations of a number of animal species. Best known are the social insects, such as ants, bees and termites, where a large number of workers contribute to the survival of the group without ever reproducing themselves. Similar phenomena have been observed in an increasing number of animal species, including social shrimps (Duffy *et al.*, 2000) and the naked mole rat (Sherman *et al.*, 1991).

The marked discrepancy between theory and observation has led me to a refinement of group selection models. It turns out that the mathematical models rely on a number of simplifications, approximations and assumptions in order to make the models mathematically tractable. This problem can be overcome by computer simulation studies. Monte Carlo simulation techniques make it possible to study models that are too complex for mathematical analysis. My simulation studies have shown that the simplifications used in mathematical analysis seriously distort the results, and that group selection can indeed be a strong force in evolution under certain conditions.

The classical models of group selection are based on geographic boundaries between groups. Most well-known is the island-model where each group lives on its own island and migration between the islands is rare (Boorman and Levitt, 1980). This model is not very realistic and it lacks a plausible explanation why groups are selectively extinguished.

I have therefore proposed a new model where groups are separated by behavioural boundaries rather than geographic boundaries. Each group has its own territory and avoids contact with neighbour groups. Group selection takes place by the mechanism of territorial warfare. A group with many brave and self-sacrificing warriors will be capable of capturing territory from neighbour groups that have more egoistic members. The egoist group will loose territory and gradually perish from lack of subsistence means. The altruist group will grow and prosper until it gets so big that it splits up in two groups. The process can then continue (Fog, 2001). A snapshot of an ongoing simulation is shown in figure 5.1.

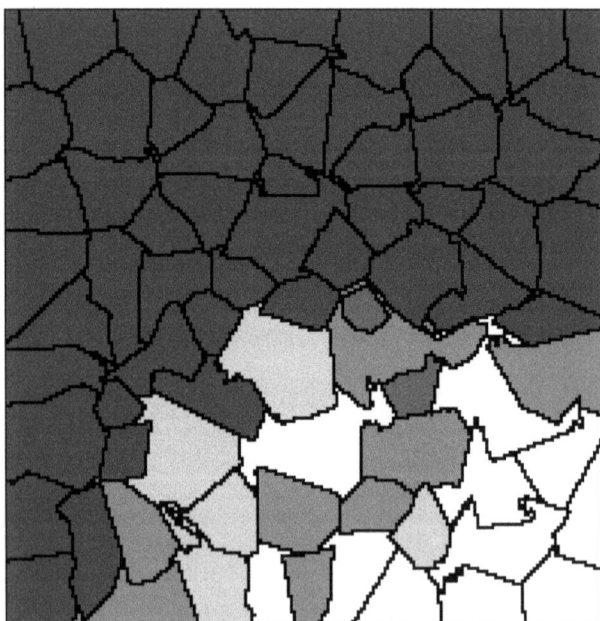

Fig. 5.1. Simulated evolution of group territories.
Lighter colours = groups with higher fraction of altruists

Computer simulations of the group territoriality model (Fog, 2000) show that this mechanism is likely to lead to

fixation of the altruism gene when the following conditions are met:

- The rate of migration or interbreeding between neighbour groups is small (< 10 % per generation).
- The group size is limited (< 1000 members).
- Few or no members of defeated groups survive.
- The altruism gene has a stronger positive effect on group fitness than the negative effect on individual fitness.
- The fitness of a group is a near-linear function of the fraction of phenotypic altruists in the group.

The pattern of group conflict behaviour that has been observed among chimpanzees in the wild (Goodall 1986) suggests that the conditions for group selection to be effective might be satisfied for these primates. My theory is that the same holds true for prehistoric man. The model of fights over group territories does indeed resemble our image of tribal warfare in prehistoric time. If the above conditions were met in prehistoric time then it is possible that this mechanism has had an important influence on human evolution.

It must be emphasized that this applies to prehistoric hunter-gatherer societies only. The first three of the above conditions are obviously not met in modern human societies and most historic cultures. It is therefore certain that any group selection mechanism that may have influenced human evolution in the past is no longer in effect.

The reason why I, as a natural scientist, present this theory at an archaeology conference is that the archaeological record may provide material that is useful for testing this model of group territoriality. A number of predictions can be made from my model, and some of these predictions can be tested against archaeological evidence. The predictions that are most relevant to archaeology can be summarized as follows:

- A strong sense of group identity should be visible in clothing, body decoration and art. These symbols of group identity should be distinct from those of neighbouring tribes.
- Tribal war should have been ubiquitous wherever there were no strong geographic boundaries preventing access to neighbour territory.
- There should be a strong cooperation within the group and a division of labour between warriors, providers, child caretakers, political and cultural (ceremonial) leaders, etc. The theory cannot say whether this division of labour is organized around gender, age or some other criteria. It is possible that everybody in the group contributes to warfare in the most critical situations.
- There may have been a well-organized justice system involving rewards for bravery in battle and punishment for defection and disloyalty.

- Defeated groups suffer a very unkind fate, and few members of a defeated group survive, if any. All members of a defeated group should either die in battle, be massacred, or die as fugitives without a territory. We can expect to find evidence of massacres including women and children.

- The victorious group can be expected to destroy any religious artefacts that the defeated group has used for seeking protection from spirits or other supernatural beings.

I regret to say that this is not a theory that paints a flattering picture of our past. It is an incredible paradox that altruism and charity cannot have evolved without merciless cruelty!

References

BOORMAN, S.A.; LEVITT, P.R. (1980) – *The Genetics of Altruism*. New York: Academic Press.

DUFFY, J.E.; MORRISON, C.L.; RIOS, R. (2000) – Multiple Origins of Eusociality among Sponge-Dwelling Shrimps (Synalpheus). *Evolution*. 54:2, p. 503-516.

FOG, A. (2000) – Simulation of Evolution in Structured Populations: The Open Source Software Package Altruist. *Biotech Software & Internet Report*. 1:5, p. 226-229.

FOG, A. (2001). *Simulation of group selection models*. http://www.agner.org/evolution/groupsel/

GOODALL, J. (1986): *The Chimpanzees of Gombe: Patterns of Behavior*. Harvard: Belknap.

SHERMAN, P.W.; JARVIS, J.U.M.; ALEXANDER, R.D., eds. (1991) – *The Biology of the Naked Mole-Rat*. Princeton University Press.

TWO FACES OF DARWIN: ON THE COMPLEMENTARITY OF EVOLUTIONARY ARCHAEOLOGY AND HUMAN BEHAVIORAL ECOLOGY

Kristen J GREMILLION

Department of Anthropology, The Ohio State University, 4034 Smith Laboratory,
174 W. 18th Avenue, Columbus, OH 43210-1106, Gremillion.1@osu.edu

Abstract: Divergent positions on the application of Darwinism lie at the heart of the divide between human behavioral ecology, with its focus on flexible responses to the environment, and evolutionary archaeology, which analyzes cultural transmission over multiple generations. Although both approaches address all four of Tinbergen's types of explanation to some degree, behavioral ecology targets functional-adaptive causes of behavior, whereas evolutionary archaeology emphasizes the evolutionary history of cultural traits. Evolutionary archaeology and archaeological behavioral ecology share a fundamentally Darwinian set of core assumptions and both must contend with the limitations of an incomplete evidential record. Both approaches have much to offer, and can coexist peacefully if their division of labor is accepted and understood.
Keywords: evolutionary archaeology, human behavioral ecology, cultural evolution, cultural transmission

Résumé: Quelques positions divergentes de l'application du Darwinisme tombent au cœur de la division entre l'écologie comportementale humaine, qui se concentre sur les réponses flexibles face à l'environnement, et l'archéologie évolutive, qui analyse la transmission culturelle à travers multiples générations. Bien que dans une certaine mesure les deux approches partagent les quatre types d'explication de Tirbergen, l'écologie comportementale vise plutôt les causes adaptative-fonctionnelle du comportement, alors que l'archéologie évolutionniste s'appuie plus sur l'histoire évolutive des traits culturels. L'archéologie évolutive et l'archéologie de l'écologie comportementale partagent un ensemble fondamentalement Darwinien de suppositions centrales et les deux doivent compter avec les limitations d'un registre d'évidences incomplet. Les deux approches ont plus que cela à offrir, et peuvent coexister pacifiquement si leur division des tâches est acceptée et comprise.
Mots-clés: archéologie évolutive, écologie humaine comportementale, évolution culturelle, transmission culturelle

> *If there are different kinds of questions to ask about the structure of the natural world, different methodologies may be needed to tackle them.*
>
> (Bowler 2003:348)

> *In the biological and social domains, "science" without "history" leaves many interesting phenomena unexplained, while "history" without "science" cannot produce an explanatory account of the past, only a listing of disconnected facts.*
>
> (Boyd and Richerson 1992:201)

> *One reason for the predominance of extreme adaptationism--but not adequate justification for its extreme versions--is the remarkable payoff, in terms of insight into function, of an uncompromising adaptationist approach.*
>
> (West-Eberhard 2003:478)

Since Dunnell's landmark publication (Dunnell 1980), methodologies for applying Darwinian evolutionary theory to the explanation of the archaeological record have proliferated. These approaches entail an explicit rejection of transformational, progressive concepts of evolution in the Spencerian mode, seeking instead to reconnect archaeology with contemporary trends in neo-Darwinism (Broughton and O'Connell 1999; Dunnell 1980, 1992; Leonard and Jones 1987; Lyman and O'Brien 1998; Neff 2000; Neff and Larson 1997; O'Brien 1996; O'Brien and Holland 1990, 1997; O'Brien and Lyman 2000, 2002; O'Brien et al. 1998; Spencer 1997; Teltser 1995). The new paradigm has elicited a variety of methodological and conceptual innovations that are currently competing for viability in the marketplace of ideas. This large pool of variation is a good sign because it ensures that if science itself is an evolutionary process

that selection will have ample material with which to work. We have reason to hope that the approaches that do the best job of accumulating accurate knowledge about the world (in this case, the human past and its present material record) can prevail.

While there is nothing wrong with pluralism, particularly in a newly developing discipline, to progress science needs to move beyond eclecticism. A coherent, unifying theoretical paradigm is a great strength and a valuable tool for accumulating knowledge. Less certain is whether such a synthesis will, or should, incorporate all of the visions of what an evolutionary archaeology should look like. I would like to believe that all of the approaches that contribute reliable scientific knowledge, and only the approaches that do so, will survive the winnowing process of debate and analysis.

I am encouraged by the current literature that something like this is actually happening, but also concerned that the pursuit of different but complementary goals in evolutionary archaeology may prove to be unnecessarily divisive. The most obvious, and explicitly argued, conflict over methods highlights differences between human behavioral ecology (the study of the adaptive design of behavior in environmental context) and evolutionary archaeology *sensu* Lyman and O'Brien (Lyman and O'Brien 1998; O'Brien and Lyman 2002) (which emphasizes the evolutionary explanation of changes in trait frequencies over time). The fierce debate between advocates of these two approaches has died down since a flurry of papers in the late 90s (Boone and Smith 1998; Broughton and O'Connell 1999; Lyman and O'Brien 1998; Neff 2000; O'Brien *et al.* 1998), and many steps have been taken to reconcile, revise, or productively combine them (e.g., Ladefoged and Graves 2000; Neff 2000). This reconciliation has been possible because some of the contrasting views on methodology are not due to fundamental disagreements about evolutionary processes but rather arise from the pursuit of different explanatory goals. (Deeper philosophical disagreements exist as well, but that is not my topic here). Because different types of explanation are not really asking the same questions, the different answers they yield need not contradict one another. I contend that far from being problematic, the coexistence of different types of explanation bodes well for the success of evolutionary archaeology as a discipline. At best, explanatory pluralism holds the promise of building more complete evolutionary explanations of change, and at worst it increases the pool of variants available to build a better science.

HUMAN BEHAVIORAL ECOLOGY AND EVOLUTIONARY ARCHAEOLOGY (NARROW SENSE) TAKE OPPOSING POSITIONS ON HOW EVOLUTIONARY THEORY SHOULD BE APPLIED IN ARCHAEOLOGICAL RESEARCH

Behavioral ecology, a specialization within the field of evolutionary ecology, has been distinguished by the use of quantitative models (many of them inspired by microeconomic research) to predict optimal behaviors in specific environments. Optimality in this context refers to the property of offering the best possible potential fitness outcome by balancing tradeoffs between competing needs (Smith and Winterhalder 1992; Stephens and Krebs 1986; Winterhalder and Smith 1992). The methodology of behavioral ecology was picked up by anthropologists and applied with some success to both contemporary and past populations (Barlow 2002; Barlow and Metcalfe 1996; Bettinger 1991; Bird and Bliege Bird 2002; Bird 1997; Bird and Bird 2000; Bliege Bird *et al.* 2002; Bliege Bird and Bird 1997; Blurton Jones 1984, 1986, 1987; Broughton 1994a, b; Broughton and Grayson 1993; Broughton 1997, 2002; Broughton and O'Connell 1999; Cannon 2003; Cannon 2000; Gardner 1992; Grayson and Delpech 1998; Grayson and Cannon 1999; Gremillion

2002; Hawkes *et al.* 1982; Hawkes *et al.* 1997; Hawkes 1990, 1991, 1993; Hill 1988; Hill *et al.* 1985, 1987; Jones and Madsen 1989; Kaplan and Hill 1992; Kaplan *et al.* 1984; Keegan 1986; Madsen 1993; Madsen and Schmitt 1998; Marlowe 2001, 2003; Nagaoka 2002; O'Connell and Hawkes 1981, 1984; Reidhead 1980; Rhode 1990; Smith 1985, 1991; Winterhalder 1980, 1983, 1996, 1999; Winterhalder *et al.* 1988; Winterhalder and Goland 1993, 1997; Winterhalder and Smith 2000). Models describing optimal solutions to survival problems such as food procurement, mate selection, group size and the like were and continue to be used to generate hypotheses to be tested against bodies of ethnographic, historical, and archaeological data.

In archaeology, human behavioral ecology (HBE) makes several explicit assumptions to guide methodology. One is the phenotypic gambit—the assumption that study of phenotypes and changes in their frequency is an adequate surrogate for tracing the underlying replicating units, whether they are packets of genetic or cultural information (Smith and Winterhalder 1992). As long as behavior varies, and its variants differ in their probability of being replicated (whether by genetic or cultural means), it can be shaped by natural selection. HBE is content to remain agnostic about the precise mechanisms of transmission. However, HBE recognizes that specific behaviors are not genetically determined, but rather assumes that decision-making capacities have been shaped by natural selection acting on genetic variation (Bird and O'Connell 2006; Broughton and O'Connell 1999; VanPool 2002). HBE asks how the choice between alternative behaviors might be a product of cognitive adaptations to seek goals, such as energy efficiency or risk reduction, that enhanced Darwinian fitness in past populations. In doing so it is explicitly adaptationist in outlook, taking optimization to be an assumption that guides research in a systematic way. This emphasis on adaptive design has been criticized as a naive faith in the power of selection to swamp sources of non-directional change and to overcome historical constraints (Boone and Smith 1998; Lyman and O'Brien 1998).

A second key assumption of HBE is that of *methodological individualism*: that the study of adaptation must be situated in the decisions made by individuals as they navigate the hazards and opportunities of everyday life (Smith and Winterhalder 1992; Winterhalder and Goland 1997). The immediate subject of interest is how human behavioral flexibility, itself a product of evolutionary processes, acts to produce optimal solutions to survival problems posed by different social and ecological conditions. While the application of HBE to archaeology aims to explain change, it does so by comparing the outcomes of human decision-making at different points in time (in contrasting environments) rather than by analyzing the evolutionary or historical processes that lead from one to the other. This sidelining of history, as perceived by some critics, calls into question the credentials of HBE as an evolutionary science.

The approach labelled "evolutionary archaeology" by its advocates is formulated in more specific terms than I think is intended by the scope of this workshop, so I am calling it evolutionary archaeology in the narrow sense (EANS). EANS derived much of its inspiration from the work of Dunnell, and has been elaborated and refined by O'Brien, Lyman, and others. A central premise of EANS is that because artifacts are elements of the human phenotype, they are appropriately studied as fossils using essentially the same methods and principles used by evolutionary biologists to explain the diversification of life forms. The pathways taken by evolution in populations of artifacts are generated by processes analogous to or identical to those that operate in the evolution of biological populations. Natural selection is mechanistic in the cultural domain, as it is in biological populations (Neff and Larson 1997; Neiman 1995). While the initial focus of EANS led to its being labeled as a selectionist approach (e.g., O'Brien and Holland 1995), its practitioners have always maintained that selection must be demonstrated and not assumed. Archaeological tests for the operation of selection include functional analysis of artifacts (performance studies) and the construction of lineage histories. Changes in trait distribution that do not meet the criteria for selection are often a product of the cultural analog of the genetic drift that occurs when lineages diverge.

The programmatic statements and applications of EANS share the conviction that a truly evolutionary archaeology must have as a primary concern the construction of historical narratives of the development of cultural lineages. These lineages are comprised of phenotypic traits, usually artifactual, that form chains of descent between donors and recipients of cultural information (Lipo et al. 2006a). Descent relations are important because only through constructing cultural phylogenies is it possible to distinguish between selection and other evolutionary processes that create patterns of convergence, parallelism, and divergence in populations. Lineage histories must be constructed on the basis of traits not subject to selection—neutral or stylistic traits— in order to assure that the resulting trees reflect shared ancestry rather than parallel adaptations of unrelated populations.

Ten years ago, EANS was attracting some criticism for its failure to generate a large body of empirically based applications, but that situation has changed. There are now a number of such studies that demonstrate the utility of EANS (Lipo et al. 1997; Lipo et al. 2006a; Neiman 1995, 1997; O'Brien et al. 1994), although it can be argued that its usefulness is constrained in some ways by programmatic orthodoxies. These include the requirement that, as in evolutionary biology, lineage construction must precede the development and testing of hypotheses to explain lineage histories in terms of evolutionary processes. Another principle of EANS that has come under fire from HBE and others is that past behavior, not being directly observable, is not an empirical entity and

therefore cannot count as data (Broughton and O'Connell 1999). While the observation that behavior does not fossilize is noncontroversial, the epistemological stance taken by EANS has been criticized as a narrow form of strict empiricism to which no working scientist could—or would wish to—conform (Boone 1997; Boone and Smith 1998). However, I do not think EANS intends the same kind of epistemological purity that characterized classical behaviorism (which strived to avoid all reference to mental processes and other unobservable phenomena as explanations of behavior). In any case, HBE has been particularly critical of the failure of EANS to address behavior more directly, and to work out its fitness implications in ecological context (Broughton and O'Connell 1999). HBE's focus on behavior and its material traces may sometimes give short shrift to tools and their evolution, though this is changing. (O'Brien and Lyman 2002), though this is changing (Bettinger et al. 2006; Ugan et al. 2003).

Despite these differences, as many have observed, HBE and EANS have considerable potential for synthesis or at least collaboration (Neff 2000; O'Brien and Lyman 2002). Both accept that archaeological research is necessary if we want to develop evolutionary explanations of cultural change. They agree that Darwinian and neo-Darwinian theory must contribute to any complete explanation of human history. Both fields have developed innovative quantitative methods for analyzing and comparing archaeological data that document changes in human behavior. Both also share the goal of explaining patterns of variation in human phenotypes over time.

HBE AND EANS EMPHASIZE DIFFERENT EXPLANATORY GOALS AND THEREFORE SHOULD EMPLOY DIFFERENT METHODS

The fact that they work at different levels of explanation can account for some, although certainly not all, of the opposition between HBE and EANS. Each approach has a distinctive set of strengths that are complementary and equally essential to developing complete evolutionary explanations. Next I want to explore these strengths as logical correlates of the explanatory goals of each approach.

Both approaches emphasize what Mayr called ultimate causality—the "why" questions about living things. In biology, questions about ultimate causes of a trait are addressed by seeking evolutionary explanations. Evolutionary explanations address both the history and the function of the trait: when it appears in different lineages, how its frequency changes, and what selective advantage is held (what Tinbergen called its survival value) (Tinbergen 1963). Likewise, in evolutionary approaches to archaeology, ultimate causality is the gold standard of explanation. Proximate causality, in contrast, refers to the immediate mechanisms that produce behavior and how those mechanisms develop during the lifetime of the organism. Proximate causality has generally been a low priority for archaeologists and evolutionary biologists, although I will argue that we can benefit from paying more attention to the psychological mechanisms behind behavior.

EANS explicitly seeks *historical explanations* for changes in the frequency of phenotypic traits (Lyman and O'Brien 1998). These changes may be the result of selection, in which case function is a relevant variable, or they may emerge because of random drift in cultural traits. So-called selectively neutral, or stylistic, traits have been of particular interest to EANS because of their utility in building phylogenetic trees. While the style/function dichotomy has been challenged (e.g., Rakita 2006), the consideration of processes analogous to genetic drift is an important counterbalance to HBE's pursuit of adaptive design explanations. Is a trend of increasing size of cooking vessels an adaptive change that makes communal meals more efficient? Or can a gradual change of this kind be more simply explained as the incremental accumulation of random variation in pot size? Simulation studies have shown that copying errors alone can create changes in trait frequency that mimic the effects of selection (Eerkens and Lipo 2005).

EANS as originally formulated has not, however, given much attention to the possibility that directional changes in trait frequency might be a product of processes other than selection or drift. Candidates for such processes have been identified as modes of cultural transmission unique to cultural, as opposed to genetic, inheritance (Boyd and Richerson 1985; Richerson and Boyd 1992). Transmission biases arise because people are not passive recipients of cultural information; instead, both individual and social learning play a central role in determining which cultural variants are successful. In a manner analogous to changes in gene frequency, cultural information can be expected to show varying patterns of spread through a population depending on the processes involved.

Patterns of this kind can be modelled quantitatively and tested against archaeological data in order to identify processes other than selection that may be operating. Ultimately, the decision rules that guide transmission biases have themselves been shaped by selection, but they have implications for the rate and magnitude of change in the archaeological record. Frequency dependent bias, for example, in which individuals tend to copy the most common trait in a set of alternatives, inhibits the spread of innovations until they reach a certain threshold. A number of studies by Bettinger and others have demonstrated that different modes of transmission can offer powerful explanations for morphological variation in artifacts such as projectile points (Bettinger and Eerkens 1999; Eerkens and Lipo 2005).

The historical emphasis of EANS is also reflected in its insistence that relations of *descent* must be established between archaeological units in order to ensure that comparisons between them measure change and not simply difference. This procedure is the one followed in paleobiology to differentiate between adaptations and traces of common descent. The analogy with biological descent is perhaps not as strong as some of these studies assume, and a number of empirical studies of cross-cultural data sets has shown that cultural lineages are sometimes strongly affected by horizontal transmission (diffusion) (Borgerhoff Mulder *et al.* 2006; Jordan and Mace 2006; MacEachern 2002).. The definition of descent could be broadened to encompass both vertical and horizontal transmission, but then descent would lack the time-transgressive character that differentiates it from transmission in general. Doing so is analogous to conflating gene flow with genetic inheritance in biology. A solution to this methodological difficulty will no doubt emerge, but its discovery calls into question the assumption that cultural descent is largely a mechanical process of faithful transmission and selective sorting. In any case, descent is less significant in archaeology than it is in biology given the greater fluidity of cultural as opposed to genetic transmission. Furthermore, the great behavioral plasticity of the human species can quickly erase traces of common descent (Futuyma 1992). EANS has answered the challenge of critics who find its concept of descent to be insufficiently precise (MacEachern 2002) by testing archaeological data sets for the effects of blending on cultural lineages, with mixed results (see papers in Lipo *et al.* (2006b)).

In contrast to EANS, HBE performs analyses of *function* without necessarily conducting a detailed study of the history of the trait in question. HBE accepts that the scientific study of natural systems requires a knowledge of their history (Winterhalder 1994:40), but maintains that it is legitimate to pursue functional explanations without establishing the line of descent linking behavioral traits over time. HBE compares material traces of behavior with predictions derived from natural selection theory and infers change from differences between assemblages that occupy different segments of the time scale and thus different environmental settings. Such a combination of functional analysis and comparison across environments is an effective way to test for the operation of a directional evolutionary process when lineage histories cannot be obtained, as when chronological information is lacking (Maxwell 1995).

As part of its concern with *function* and its implications for selection, HBE places emphasis on the *environmental context* of behavior. The application of foraging theory and other sets of optimization models to archaeological data sets relies heavily on detailed reconstructions of past environments. For this reason, HBE researchers continue to be puzzled by EANS' lack of interest in paleoecology. While the functional, or performance, characteristics of artifacts are a key tool of EANS, they are not always measured relative to a range of environmental conditions (for example, pottery used to cook different types of seeds or dart points against different prey animals). EANS advocates the measurement of physical and chemical properties of artifacts that are invariant across time and space, regardless of environmental context (Lyman and O'Brien 1998).

ANALYSIS OF PROXIMATE CAUSALITY CAN SUPPORT INFERENCES OF ULTIMATE CAUSE

Finally, I offer some food for thought regarding proximate causes: the mechanisms that motivate behavior and how they develop. The psychological mechanisms behind behavior may seem largely irrelevant to evolutionary studies in archaeology; inferences about what behaviors formed the archaeological record is difficult enough without trying to identify their internal causes. However, there are good reasons for HBE to pay more attention to behavioral mechanisms. The most important of these is that the inference that a given behavior is adaptive rests on the assumption that cognitive mechanisms, such as decision rules, perception, and information processing by the brain generally, have been shaped by natural selection in the past. This is a reasonable assumption, but one that warrants further testing. For at least the past 40,000 years or so, human cognition (or at least the basic structure of the brain) has probably been sufficiently constant to support a uniformitarian assumption. This means that experimental results from cognitive psychology can yield relevant details about how humans process information from their environment and make decisions accordingly. I am not suggesting that particular decisions are not conditioned by the cultural environment or that they are in any way determined by brain biology. However, insights from studies of economic decision making (for example) are likely to be useful to HBE in at least two ways. First, they function as tests of the assumption that people manage tradeoffs between competing needs by optimizing. The diet breadth model, for example, assumes that a forager's goal is to choose a diet breadth that provides the best average rate of energy acquisition. The solution proposed by the model balances the tradeoff between the quality and density of resources. It would be helpful to revisit some of the empirical support for the optimization assumption, partly as a response to critics who accuse HBE of ultra-adaptationism. However, experimental studies of decision-making can also stimulate the development of new models that might be tested archaeologically. The potential of this approach is suggested by some recent considerations of the spread of innovations and its archaeological correlates (Bettinger and Eerkens 1999; Bettinger et al. 2006; Fitzhugh 2001).

Like behavioral mechanisms, the development of behavior during the lifetime of the organism leaves little archaeological trace. For this reason, developmental causation might seem to be of little relevance to the understanding of historical changes in trait distribution.

Recent research in evolutionary developmental biology has shown with increasing clarity that the genotype-phenotype correlation, while strong enough to allow selection to operate, is much less straightforward than previously thought. Large phenotypic differences may arise from small genetic changes in genes that regulate development. In similar fashion, genes that underlie plastic responses to environment are not subject to selection until those responses are triggered and become part of the phenotype (Bateson 2004; West-Eberhard 2003). For example, certain genetically-based motor skills are much more likely to be expressed, and selected for or against, in an environment in which tool-making is widely practiced and taught. The emerging field of evolutionary developmental biology is proving to be a rich source of insights into the relationships between genetic and behavioral variation. Some of these insights may yield testable hypotheses for the evolutionary archaeologist.

I have tried to show that the ecological-functional approach of HBE and the more historical emphasis of EANS are for the most part complementary—representing the two faces of Darwin in my title. Darwin the naturalist is the ancestor of HBE. This is the Darwin who marveled at the coevolution of highly specialized orchid flowers with their insect pollinators and saw in the feeding behavior and morphology of Galapagos finches diverse adaptations to local food sources (Darwin 1859). While he was cautious about applying his insights to human behavior, Darwin's observations of the organism in its environment provided much of the empirical support for the theory of evolution by natural selection. But there is also the Darwin who saw adaptation as the culmination of a historical process of relentless culling of non-beneficial variants and who recognized the branching character of evolutionary divergence from a common ancestor. Historical depth and evolutionary process were also at the heart of the *Origin*. There is no reason not to respect and draw on both aspects of Darwinism while continuing to refine our understanding of evolutionary processes as they operate on human behavioral variation. History is not an alternative to adaptation, but rather a complement to it (Borgerhoff Mulder et al. 2006). Given today's broad array of sophisticated techniques for observing the archaeological record and analyzing archaeological data, some degree of specialization is necessary and appropriate.

Perhaps dividing the intellectual labor in recognition of different types of explanation will make our task easier and yield more complete explanations of human history than we would otherwise obtain.

References

BARLOW, K.R. (2002) Predicting Maize Agriculture among the Fremont: An Economic Comparison of Farming and Foraging in the American Southwest. *American Antiquity* 67:65-88.

BARLOW, K.R. and D. METCALFE (1996) Plant Utility Indices: Two Great Basin Examples. *Journal of Archaeological Science* 23:351-371.

BATESON, PATRICK (2004) The Active Role of Behavior in Evolution. *Biology and Philosophy* 19:283-298.

BETTINGER, ROBERT L. (1991) *Hunter-Gatherers: Archaeological and Evolutionary Theory*. Plenum, New York.

BETTINGER, ROBERT L. and JELMER EERKENS (1999) Point Typologies, Cultural Transmission, and the Spread of Bow-and-Arrow Technology in the Prehistoric Great Basin. *American Antiquity* 64:231-242.

BETTINGER, ROBERT L., BRUCE WINTERHALDER and RICHARD MCELREATH (2006) A Simple Model of Technological Intensification. *Journal of Archaeological Science* 33:538-545.

BIRD, DOUGLAS and R.L. BLIEGE BIRD (2002) Children on the Reef: Slow Learning or Strategic Foraging? *Human Nature* 13:269-298.

BIRD, DOUGLAS and JAMES F. O'CONNELL (2006) Behavioral Ecology and Archaeology. *Journal of Archaeological Research* 14:143-188.

BIRD, DOUGLAS W. (1997) Behavioral Ecology and the Archeological Consequences of Central Place Foraging among the Meriam. In *Rediscovering Darwin: Evolutionary Theory and Archeological Explanations*, edited by C.M. Barton and G.A. Clark, pp. 291-308. 7. Archeological Papers of the American Anthropological Association. American Anthropological Association, Arlington, Virginia.

BIRD, DOUGLAS W. and REBECCA BLIEGE BIRD (2000) The Ethnoarchaeology of Juvenile Foragers: Shellfishing Strategies among Meriam Children. *Journal of Anthropological Archaeology* 19:461-476.

BLIEGE BIRD, REBECCA, DOUGLAS BIRD, E.A. SMITH and GEOFFREY C. KUSHNIK (2002) Risk and Reciprocity in Meriam Food Sharing. *Evolution and Human Behavior* 23:297-321.

BLIEGE BIRD, REBECCA and DOUGLAS W. BIRD (1997) Delayed Reciprocity and Tolerated Theft: The Behavioral Ecology of Food-Sharing Strategies. *Current Anthropology* 38:49-78.

BLURTON JONES, NICHOLAS G. (1984) A Selfish Origin for Human Food Sharing: Tolerated Theft. *Ethology and Sociobiology* 5:1-3.

BLURTON JONES, NICHOLAS G. (1986) Bushman Birth Spacing: A Test for Optimal Interbirth Intervals. *Ethology and Sociobiology* 7:91-105.

BLURTON JONES, NICHOLAS G (1987) Bushman Birth Spacing: Direct Tests of Some Simple Predictions. *Ethology and Sociobiology* 8:183-204.

BOONE, JAMES L. (1997) Review of Darwinian Archaeologies, Edited by Herbert Donald Graham Maschner. *Evolution and Human Behavior* 18:439-442.

BOONE, JAMES L. and ERIC ALDEN SMITH (1998) Is It Evolution Yet? A Critique of Evolutionary Archaeology. *Current Anthropology* 39:S141-S173.

BORGERHOFF MULDER, MONIQUE, CHARLES L. NUNN and MARY C. TOWNER (2006) Cultural Macroevolution and the Transmission of Traits. *Evolutionary Anthropology* 15:52-64.

BOWLER, PETER J. (2003) *Evolution: The History of an Idea*. Third ed. University of California Press, Berkeley.

BOYD, ROBERT and PETER J. RICHERSON (1985) *Culture and the Evolutionary Process*. University of Chicago Press, Chicago.

BOYD, ROBERT and PETER J. RICHERSON (1992) How Microevolutionary Processes Give Rise to History. In *History and Evolution*, edited by D.H. Nitecki and M.V. Nitecki, pp. 179-209. State University of New York Press, Albany.

BROUGHTON, J.M. (1994a) Declines in Mammalian Foraging Efficiency During the Late Holocene, San Francisco Bay, California. *Journal of Anthropological Archaeology* 13:371-401.

BROUGHTON, J.M. (1994b) Late Holocene Resource Intensification in the Sacramento Valley: The Archaeological Vertebrate Evidence. *Journal of Archaeological Science* 21:501-514.

BROUGHTON, J.M. and D.K. GRAYSON (1993) Diet Breadth, Numic Expansion, and the White Mountain Faunas. *Journal of Archaeological Science* 20:331-336.

BROUGHTON, J.M. and D.K. GRAYSON (1997) Widening Diet Breadth, Declining Foraging Efficiency, and Prehistoric Harvest Pressure: Icthyofaunal Evidence from the Emeryville Shell Mound, California. *Antiquity* 71:845-862.

BROUGHTON, J.M. and D.K. GRAYSON (2002) Prey Spatial Structure and Behavior Affect Archaeological Tests of Optimal Foraging Models: Examples from the Emeryville Shellmound Vertebrate Fauna. *World Archaeology* 34:60-83.

BROUGHTON, JACK M. and JAMES F. O'CONNELL (1999) On Evolutionary Ecology, Selectionist Archaeology, and Behavioral Archaeology. *American Antiquity* 64:153-165.

CANNON, MICHAEL (2003) A Model of Central Place Forager Prey Choice and an Application to Faunal Remains from the Mimbres Valley, New Mexico. *Journal of Anthropological Archaeology* 22:1-25.

CANNON, MICHAEL D. (2000) Large Mammal Relative Abundance in Pithouse and Pueblo Period Archaeofaunas from Soutwestern New Mexico: Resource Depression among the Mimbres-Mogollon? *Journal of Anthropological Archaeology* 19:317-347.

DARWIN, CHARLES (1859) *The Origin of Species.* Random House, New York.

DUNNELL, ROBERT C. (1980) Evolutionary Theory and Archaeology. *Advances in Archaeological Method and Theory* 3:35-91.

DUNNELL, ROBERT (1992) Archaeology and Evolutionary Science. In *Quandaries and Quests: Visions of Archaeology's Future*, edited by L. Wandsnider, pp. 209-224. Center for Archaeological Investigation, Southern Illinois University, Carbondale.

EERKENS, JELMER W. and CARL P. LIPO (2005) Cultural Transmission, Copying Errors, and the Generation of Variation in Material Culture and the Archaeological Record. *Journal of Anthropological Archaeology* 24:316-334.

FITZHUGH, BEN (2001) Risk and Invention in Human Technological Evolution. *Journal of Anthropological Archaeology* 20:125-167.

FUTUYMA, DOUGLAS (1992) History and Evolutionary Processes. In *History and Evolution*, edited by D.H. Nitecki and M.V. Nitecki, pp. 104-130. State University of New York Press, Albany.

GARDNER, PAUL S. (1992) *Diet Optimization Models and Prehistoric Subsistence Change in the Eastern Woodlands.* Unpublished Ph.D. Dissertation, Department of Anthropology, University of North Carolina.

GRAYSON, D.K. and F. DELPECH (1998) Changing Diet Breadth in the Early Upper Palaeolithic of Southwestern France. *Journal of Archaeological Science* 25:1119-1130.

GRAYSON, DONALD K. and MICHAEL D. CANNON (1999) Human Paleoecology and Foraging Theory in the Great Basin. In *Models for the Millennium: The Current Status of Great Basin Anthropological Research*, edited by C. Beck, pp. 141-151. University of Utah Press, Salt Lake City.

GREMILLION, KRISTEN J. (2002) Central Place Foraging and the Origins of Food Production in Eastern Kentucky. In *Foraging Theory and Thetransition to Agriculture*, edited by D.J. Kennett and B. Winterhalder. Smithsonian Institution Press, Washington, D.C.

HAWKES, K., KIM HILL and JAMES F. O'CONNELL (1982) Why Hunters Gather: Optimal Foraging and the Ache of Eastern Paraguay. *American Ethnologist* 9:379-398.

HAWKES, K., J.F. O'CONNELL and L. ROGERS (1997) The Behavioral Ecology of Modern Hunter-Gatherers and Human Evolution. *Trends in Ecology and Evolution* 12:29-32.

HAWKES, KRISTEN (1990) Why Do Men Hunt? Some Benefits for Risky Choices. In *Risk and Uncertainty in Tribal and Peasant Economies*, edited by E. Cashdan, pp. 145-166. Westview, Boulder, Colorado.

HAWKES, KRISTEN (1991) Showing Off: Tests of Another Hypothesis About Men's Foraging Goals. *Ethology and Sociobiology* 11:29-54.

HAWKES, KRISTEN (1993) Why Hunter-Gatherers Work: An Ancient Version of the Problem of Public Goods. *Current Anthropology* 34:341-361.

HILL, K. (1988) Macronutrient Modifications of Optimal Foraging Theory: An Approach Using Indifference Curves Applies to Some Modern Foragers. *Human Ecology* 16:157-197.

HILL, KIM, HILLARD KAPLAN, KRISTEN HAWKES and ANA MAGDALENA HURTADO (1985) Men's Time Allocation to Subsistence Work among the Ache of Eastern Paraguay. *Human Ecology* 13:29-47.

HILL, KIM, HILLARD KAPLAN, KRISTEN HAWKES and ANA MAGDALENA HURTADO (1987) Foraging Decisions among Ache Hunter-Gatherers: New Data and Implications for Optimal Foraging Models. *Ethology and Sociobiology* 8:1-36.

JONES, K.T. and D.B. MADSEN (1989) Calculating the Cost of Resource Transportation: A Great Basin Example. *Current Anthropology* 30:529-534. Jordan, Peter and Thomas Mace

JONES, K.T. and D.B. MADSEN (2006) Tracking Culture-Historical Lineages: Can "Descent with Modification" Be Linked To "Association by Descent"? In *Mapping Our Ancestors: Phylogenetic Approaches in Archaeology and History*, edited by C.P. Lipo, M. Collard, M.J. O'Brien and S. J. Shennan, pp. 149-167. Aldine Transaction, New Brunswick, U. S. A.

KAPLAN, HILLARD and KIM HILL (1992) The Evolutionary Ecology of Food Acquisition. In *Evolutionary Ecology and Human Behavior*, edited by E.A. Smith and B. Winterhalder, pp. 167-202. Aldine de Gruyter, New York.

KAPLAN, HILLARD, KIM HILL, KRISTEN HAWKES and ANA MAGDALENA HURTADO (1984) Food Sharing among Ache Hunter-Gatherers of Eastern Paraguay. *Current Anthropology* 25:113-115.

KEEGAN, WILLIAM F. (1986) The Optimal Foraging Analysis of Horticultural Production. *American Anthropologist* 88:92-107.

LADEFOGED, THEGN N. and MICHAEL W. GRAVES (2000) Evolutionary Theory and the Historical Development of Dry-Land Agriculture in North Kohala, Hawai'i. *American Antiquity* 65:423-448.

LEONARD, ROBERT D. and G.T. JONES (1987) Elements of an Inclusive Evolutionary Model in Archaeology. *Journal of Anthropological Archaeology* 6:199-219.

LIPO, CARL P., MARK E. MADSEN, ROBERT C. DUNNELL and TIM HUNT (1997) Population Structure, Cultural Transmission, and Frequency Seriation. *Journal of AnthropoloGical Archaeology* 16:301-333.

LIPO, CARL P., MICHAEL J. O'BRIEN, MARK COLLARD and STEPHEN J. SHENNAN (2006a) Cultural Phylogenies and Explanation: Why Historical Methods Matter. In *Mapping Our Ancestors: Phylogenetic Approaches in Archaeology and History*, edited by C.P. Lipo, M.J. O'Brien, M. Collard and S. J. Shennan, pp. 3-16. Aldine Transaction, New Brunswick, U. S. A.

LIPO, CARL P., MICHAEL J. O'BRIEN, MARK COLLARD and STEPHEN J. SHENNAN (2006b) *Mapping Our Ancestors: Phylogenetic Approaches in Archaeology and History*. Aldine Transaction, New Brunswick, USA.

LYMAN, R. LEE and MICHAEL J. O'BRIEN (1998) The Goals of Evolutionary Archaeology: History and Explanation. *Current Anthropology* 39:615-652.

MacEachern, Scott

LYMAN, R. LEE and MICHAEL J. O'BRIEN (2002) Descent. In *Darwin and Archaeology: A Handbook of Key Concepts* edited by J.P. Hart and J.E. Terrell, pp. 125-141. Bergin and Garvey, Westport, Connecticut.

MADSEN, D.B. (1993) Testing Diet Breadth Models: Examining Adaptive Change in the Late Prehistoric Great Basin. *Journal of Archaeological Science* 20:321-330.

MADSEN, D.B. and D.N. SCHMITT (1998) Mass Collecting and the Diet Breadth Model: A Great Basin Example. *Journal of Archaeological Science* 25:445-456.

MARLOWE, FRANK (2001) Male Contribution to Diet and Female Reproductive Success among Foragers. *Current Anthropology* 42:755-759.

MARLOWE, FRANK (2003) A Critical Period for Provisioning by Hadza Men: Implications for Pair Bonding. *Evolution and Human Behavior* 24:217–229.

MAXWELL, TIMOTHY D. (1995) The Use of Comparative and Engineering Analyses in the Study of Prehistoric Agriculture. In *Evolutionary Archaeology: Methodological Issues*, edited by P.A. Teltser, pp. 113-128. University of Arizona Press, Tucson.

NAGAOKA, LISA (2002) Explaining Subsistence Change in Southern New Zealand Using Foraging Theory Models. *World Archaeology* 34:84-102.

NEFF, HECTOR (2000) On Evolutionary Ecology and Evolutionary Archaeology: Some Common Ground? *Current Anthropology* 41:427-428.

NEFF, HECTOR and DANIEL O. LARSON (1997) Methodology of Comparison in Evolutionary Archaeology. In *Rediscovering Darwin: Evolutionary Theory and Archeological Explanations*, edited by C.M. Barton and G.A. Clark, pp. 75-94. American Anthropological Association, Arlington, Virginia.

NEIMAN, F.D. (1995) Stylistic Variation in Evolutionary Perspective: Implications for Middle Woodland Ceramic Diversity. *American Antiquity* 60:7-36.

NEIMAN, F.D. (1997) Conspicuous Consumption as Wasteful Advertising: A Darwinian Perspective on Spatial Patterns in Classic Maya Terminal Monument Dates. In *Rediscovering Darwin: Evolutionary Theory and Archeological Explanations*, edited by C.M. Barton and G.A. Clark, pp. 267-290. American Anthropological Association, Washington, D.C.

O'BRIEN, MICHAEL J. (editor) (1996) *Evolutionary Archaeology: Theory and Application*. University of Utah Press, Salt Lake City.

O'BRIEN, MICHAEL J. and THOMAS D. HOLLAND (1990) Variation, Selection, and the Archaeological Record. *Archaeological Method and Theory* 2:31-79.

O'BRIEN, MICHAEL J. and THOMAS D. HOLLAND (1995) The Nature and Premise of a Selection-Based Archaeology. In *Evolutionary Archaeology: Methodological Issues*, edited by P.A. Teltser, pp. 175-200. University of Arizona Press, Tucson.

O'BRIEN, MICHAEL J. and THOMAS D. HOLLAND (1997) The Role of Adaptation in Archaeological Explanation. *American Antiquity* 57:3-59.

O'BRIEN, MICHAEL J., THOMAS D. HOLLAND, R.J. HOARD and G.L. FOX (1994) Evolutionary Implications of Design and Performance Characteristics of Prehistoric Pottery. *Journal of Archaeological Method and Theory* 1:259-304

O'BRIEN, MICHAEL J. and R. LEE LYMAN (2000) *Applying Evolutionary Archaeology: A Systematic Approach*. Kluwer Academic/Plenum, New York.

O'BRIEN, MICHAEL J. and R. LEE LYMAN (2002) Evolutionary Archeology: Current Status and Future Prospects. *Evolutionary Anthropology* 11:26-36.

O'BRIEN, MICHAEL J., R. LEE LYMAN and ROBERT D. LEONARD (1998) Basic Incompatibilities between Evolutionary and Behavioral Archaeology. *American Antiquity* 63:485-498.

O'CONNELL, J.F. and K. HAWKES (1981) Alyawara Plant Use and Optimal Foraging Theory. In *Hunter-Gatherer Foraging Strategies: Ethnographic and Archaeological Analyses*, edited by B. Winterhalder and E.A. Smith, pp. 99-125. University of Chicago Press, Chicago.

O'CONNELL, J.F. and K. HAWKES (1984) Food Choice and Foraging Sites among the Alyawara. *Journal of Anthropological Research* 40:504-535.

RAKITA, GORDON F.M. (2006) Phylogenetic Techniques and Methodological Lessons from Bioarchaeology. In *Mapping Our Ancestors: Phylogenetic Approaches in Archaeology and History*, edited by C.P. Lipo, M.J. O'Brien, M. Collard and S.J. Shennan, pp. 119-129. Aldine Transaction, New Brunswick, U.S.A.

REIDHEAD, V.A. (1980) The Economics of Subsistence Change: Test of an Optimization Model, edited by T.K. Earle and A.L. Christenson, pp. 141-186. Academic Press, New York.

RHODE, D. (1990) On Transportation Costs of Great Basin Resources: An Assessment of the Jones-Madsen Model. *Current Anthropology* 31:413-419.

RICHERSON, PETER J. and ROBERT BOYD (1992) Cultural Inheritance and Evolutionary Ecology. In *Evolutionary Ecology and Human Behavior*, edited by E.A. Smith and B. Winterhalder, pp. 61-92. Aldine de Gruyter, New York.

SMITH, ERIC ALDEN (1985) Inuit Foraging Groups. *Ethology and Sociobiology* 6:27-47.

SMITH, ERIC ALDEN (1991) *Inujjuamit Foraging Strategies: Evolutionary Ecology of an Arctic Hunting Economy*. Aldine de Gruyter, New York.

SMITH, ERIC ALDEN and BRUCE WINTERHALDER (1992) Natural Selection and Decision Making: Some Fundamental Principles. In *Evolutionary Ecology and Human Behavior*, edited by E.A. Smith and B. Winterhalder, pp. 25-60. Aldine de Gruyter, New York.

SPENCER, CHARLES S. (1997) Evolutionary Approaches in Archaeology. *Journal of Archaeological Research* 5:209-264.

STEPHENS, DAVID W. and JOHN R. KREBS (1986) *Foraging Theory*. Princeton University Press, Princeton.

TELTSER, PATRICE A. (editor) (1995) *Evolutionary Archaeology: Methodological Issues*. University of Arizona Press, Tucson.

TINBERGEN, N. (1963) On the Aims and Methods of Ethology. *Zeitschrift fur Tierpsychologie* 20:410-433.

UGAN, ANDREW, JASON BRIGHT and ALAN ROGERS (2003) When Is Technology Worth the Trouble? *Journal of Archaeological Science* 30:1315-1329.

VANPOOL, TODD L. (2002) Adaptation. In *Darwin and Archaeology: A Handbook of Key Concepts*, edited by J.P. Hart and J.E. Terrell, pp. 15-28. Bergin and Garvey, Westport, Connecticut.

WEST-EBERHARD, MARY JANE (2003) *Developmental Plasticity and Evolution*. Oxford, Oxford.

WINTERHALDER, BRUCE (1980) Environmental Analysis in Human Evolution and Adaptation Research. *Human Ecology* 8:135-170.

WINTERHALDER, BRUCE (1983) Opportunity-Cost Foraging Models for Stationary and Mobile Predators. *American Naturalist* 122:73-84.

WINTERHALDER, BRUCE (1994) Concepts in Historical Ecology: The View from Evolutionary Theory. In *Historical Ecology: Cultural Knowledge and Changing Landscapes*, edited by C.L. Crumley, pp. 17-42. School of American Research Press, Santa Fe.

WINTERHALDER, BRUCE (1996) Social Foraging and the Behavioral Ecology of Intragroup Resource Transfer. *Evolutionary Anthropology* 5:46-57.

WINTERHALDER, BRUCE (1999) Intra-Group Resource Transfers: Comparative Evidence, Models, and Implications for Human Evolution. In *Meat Eating and Human Evolution*, edited by C.B. Stanford and H.T. Bunn. Oxford University Press, Oxford.

WINTERHALDER, BRUCE, WILLIAM BAILLARGEON, FRANCESCA CAPPALLETTO, JR. I. RANDOLPH DANIEL and CHRIS PRESCOTT (1988) The Populations Ecology of Hunter-Gatherers and Their Prey. *Journal of Anthropological Archaeology* 7:289-328.

WINTERHALDER, BRUCE and CAROL GOLAND (1993) On Population, Foraging Efficiency, and Plant Domestication. *Current Anthropology* 34:710-715.

WINTERHALDER, BRUCE and CAROL GOLAND (1997) An Evolutionary Ecology Perspective on Diet Choice, Risk, and Plant Domestication. In *People, Plants, and Landscapes: Studies in Paleoethnobotany*, edited by K.J. Gremillion, pp. 123-160. University of Alabama Press, Tuscaloosa.

WINTERHALDER, BRUCE and ERIC ALDEN SMITH (1992) Evolutionary Ecology and the Social Sciences. In *Evolutionary Ecology and Human Behavior*, edited by E.A. Smith and B. Winterhalder, pp. 3-24. Aldine de Gruyter, New York.

WINTERHALDER, BRUCE and ERIC ALDEN SMITH (2000) Analyzing Adaptive Strategies: Human Behavioral Ecology at Twenty-Five. *Evolutionary Anthropology* 9:51-72.

THE STUDY OF THE ARCHAEOLOGICAL RECORD OF SANTA ROSA DE LOS PASTOS GRANDES, PUNA OF SALTA, ARGENTINA, FROM AN INCLUSIVE EVOLUTIONARY PERSPECTIVE

Gabriel LÓPEZ

Grupo de Investigación Cultura, Comportamiento y Evolución Humana (GICCEH), Sección Arqueología, Universidad de Buenos Aires. CONICET, 25 de Mayo 217 3º Piso, Buenos Aires, (1002) ARG

Abstract: Darwinian evolutionism presents different theoretical perspectives that can be applied to the study of the archaeological record. In this work, the evolutionary ecology is particularly emphasized, for being a theoretical framework that allows to analyze human behavior from formal models, and it is applicable to the interpretation of the archaeological evidence. The case of study corresponds to an area of the Highlands of the Northwest of Argentina, Puna of Salta, Pastos Grandes, which average altitude is superior to 4000 meters above the sea level and the risk of unpredictable droughts is very high. The archaeological record is represented by evidence of high and low density, as much in surface as in layer. From models of human behavioral ecology , hypotheses of optimization and risk management are proposed for the study of archaeological materials, in special, archaeofaunas and lithics. Nevertheless, the social aspects that articulate human behaviors are not left aside. These are considered from hypotheses derived from the mechanisms of the Theory of the Cultural Transmission. In this sense, it is emphasized the importance of including different darwinian perspectives that allow to understand the variability of human behavior from the archaeological record and different lines of evidence for the empirical application of this theoretical framework.
Keywords: Inclusive Evolutionary Perspective- Archaeological Record- Puna of Salta

Résumé: L'évolutionnisme Darwinien présente différentes perspectives théoriques qui peuvent être appliquées à l'étude du registre archéologique. Dans ce travail, l'écologie évolutive est particulièrement mise en valeur, pour être une approche théorique qui permet d'analyser les comportements humains à partir de modèles formels applicables à l'interprétation de l'évidence archéologique. Le cas étudié correspond à une région des hautes terres du Nord-ouest de l'Argentine, Puna de Salta, Pastos Grandes, où l'altitude moyenne est supérieure à 4000 m snm. et où le risque de sécheresses imprévisibles est très élevé. Le registre archéologique présente des densités de matériel hautes et basses, aussi bien en surface qu'en stratigraphie.
A partir des modèles de l'écologie des comportements humains, des hypothèses d'optimisation et de gestion du risque sont proposées pour l'étude du matériel archéologique, particulièrement des restes de faune et des pièces lithiques. Néanmoins, les aspects sociaux articulant les comportements humains ne sont pas mis de côté. Ces aspects sont pensés à partir d'hypothèses provenant des mécanismes de la Théorie de la Transmission Culturelle. Dans ce sens, la complémentarité de différentes perspectives darwiniennes est soulignée dans la mesure qu'elle permet la variabilité du comportement humain à partir du registre archéologique et des différentes lignes d'évidences pour l'application empirique de cette approche théorique.
Mots clés: perspective evolutive complementaire- registre archeologique- puna de Salta

INTRODUCTION

In the last 20 years, the darwinian evolutionism has became an important theoretical framework in Archaeology.

In this paper, I will develop an evolutionary ecology perspective (Boone and Smith 1998, Smith 1992, Smith and Winterhalder 1992) also regarding the mechanisms of cultural transmission in the human adaptation (Richerson and Boyd 1992). I consider important the unification of different theoretical perspectives of the darwinian evolucionism. Although each theoretical framework has its own agenda of studies and specific objectives, it is possible to achieve a complementariness. Therefore, the agenda of studies of the evolutionary ecology could be extended and include the social mechanism that influence our specie. From this perspective, the aim of this paper is to make a contribution to the diversity of the human occupations in an environment of high risk, the Salta's Puna, Argentina.

The area of study is the basin of Pastos Grandes, which is localized over 4000 meters above the sea level. The environment of the Puna is characterized for being highly risky due to the fact that the weather fluctuations are unpredictable.

In the area it is possible to determine three different geoenvironmental or geoecological sectors: the principal vega, the valleys or gorges and the salt flat.

DARWINIAN EVOLUCIONISM: MODELS, UNITS AND HYPOTHESIS

The evolutionary ecology use the hypothetic-deductive method of investigation (Smith 1992). Thus, from formal model it is possible to propose hypothesis to compare them with the empirical cases. In the archaeological record, the decision making can't be seen in an ecological time because generally we have an average of behaviors of different individuals along the time. Therefore, it is important to consider units that allow to apply this models in archaeological scale. The construction of units as measuring tools of the variation is fundamental for the archaeologists (Muscio and López ms). In this point it is essential the evaluation of the reliability and validity of the analysis' units (Ramenofsky and Steffen 1998). The

Table 7.1: Radiocarbonic dates of Alero Cuevas site

Site	Laboratory	Radiocarbon date BP
Alero Cuevas AC/C2/X	AA66544 NSF – Arizona AMS Laboratory	643 ± 35
Alero Cuevas AC/C1-C1	LP- 1671	2020 ± 60
Alero Cuevas AC/C2-F2	LP- 1655	4210 ± 70
Alero Cuevas AC/C2/ F3	LP- 1759	6510 ± 80
Alero Cuevas AC/ C2-F4	AA 71135 NSF – Arizona AMS laboratory	8504 ± 52
Alero CuevasAC-C2- F4	AA 71136 NSF – Arizona AMS laboratory	8838 ± 52
Alero Cuevas AC- C4- F4	LP- 1736	9650 ± 100

former is the precision and exactitude of the units, and the latter is the empiric or abstract correspondence of the units according to the aims of the investigation.

The units of analysis are theoretical and empirical (Muscio and López ms). The theoretical units are used exclusively as concepts and the empirical units, although ideational entities, are to refer to physical things. For example, the theoretical units are *metapopulation*, *local population* and *ocupation*. The empirical units are *areal archaeological structure*, *sectorial archaeological land-scape*, *component*, *site fraction*, *specimen* and *attribute* (see definitions in Muscio and López ms). Each unit represents archaeological evidence aggregates in different levels of inclusiveness. In their insides, they can be separated into assemblages and sub-assemblages.

Theoretically speaking, it is expected that the Pastos Grandes' basin will show the mechanism of decision making and cultural transmission of the local population in its own habitat along the history of its occupation. In this sense, from areal archaeological structure taken as a sample of a continuous distribution of archaeological evidence (Pastos Grandes's basin) I expect to have access to information in a population level. However sometimes the information at this level might come from a particular sector of the area, or from a site that indicates a sequence of occupation (see the case of Alero Cuevas). Each occupation will leave an empirical signal in the component, defined as the distribution of materials temporally distinguished from other distributions. For this matter, it is necessary tools that measure the time, radiocarbonic dates, stratigraphy or historic types of artifacts.

The analysis begins from the minor level of inclusiveness (specimens and attributes) and it reaches higher levels (component, areal archaeological structure).

At hypothetic level, in the analyzed components, I suggest that the human groups of the area of study developed a general strategy to minimize the risk in order to adapt to the environment. This general strategy would be characterized by the conjunction of different strategies aimed at the diversification for minimize the risk and also

obtain high returns. This will be reflected on the use of faunal resources through the diversification in the consume of camelids (resources of the highest return in the Puna). Also I expect to find a conjunction of strategies of high and low cost in the production of lithic tools for minimize the risk (Bousman 1993). The role of cultural transmission would allow a faster and more effective adaptation to the environment (Richerson and Boyd 1992), from bias in the confection of certain lithic technologies and consume of resources.

THE ARCHAEOLOGICAL RECORD OF PASTOS GRANDES: SPATIAL DISTRIBUTION AND TEMPORALITY

The archaeological prospections were made in the three geoenvironmental sectors of the area: the vega, the valleys o gorges and the salt flat.

The main archaeological site regarded in this paper is Alero Cuevas. This is a eaves multicomponent with occupations from the early Holocene. At this moment, the radiocarbonic dates in Alero Cuevas are: 9650±100 AP, 8838±52 AP, 8504±52 AP, 6510±80 AP, 4210±70 AP, 2020±60 y 643±35 AP (Table 7.1, see also López ms). Therefore, there is a sequence of human occupations along of the Holocene: in early Holocene, Middle Holocene and later Holocene. The site is located in the Quebrada de la Cuevas (Gorge). This eaves have 19.3 meters of extension and 8.7 m. until the dripping line in its deepest part (López 2005). The excavation was made following the natural layers of the eaves.

In the edge of the salt flat, it is located another archaeological site which is analyzed in this paper. The site La Hoyada is characterized by the presence in surface of a lithic material's concentration. In this site it was also made a transect to obtain a reliable statistic sample.

In order to assess the spacial artifactual distribution of the area, this paper focuses these sites (Alero Cuevas and La Hoyada) as well as low density surface artifact distributions.

Regarding the temporality of the sites presented, I will focus in the components corresponding to two different moments of the later Holocene: one dated in 2020 BP and other in 4210 BP. These dates came from two layers of Alero Cuevas. The date of ca. 4200 BP is related to other sites of the Puna where hunter gatherers got a reduced residential mobility while possibly practicing activities connected with the domestication of the camelids (Aschero and Yacobaccio 1999, Yacobaccio 2001). On the other hand, the date of ca. 2000 BP corresponds to a moment of strong develop of the pastoralism (Yacobaccio *et al.* 1997).

METHODOLOGY AND RESULTS

Firstly, the methodology consisted in the delimitation of the analysis' units. I considered for this matter the different empirical units mentioned before. The principal methodological aim is to achieve an approach to population level from the more inclusive units (e. g. the areal archaeological structure). For this reason, I established different assemblages as a consequence of the chronology association (radiocarbon dates) and for the attributes shared between specimens. The radiocarbonic dates allow to distinguish different components. These empirical units would be theoretically corresponding with the concept of occupation. On the other hand, the attributes shared between some artifacts (see further on the unifacial lanceolate artifacts), recuperated in different sectors and components, contribute to analyze patters of cultural transmission in the population level.

For each kind of record I used a particular methodology. In the case of the lithic material I principally followed the proposal of Aschero (1983) and for the archaeofaunal material, the Mengoni Goñalons one (1999).

LITHIC ANALYSIS

The attributes considered in each specimen of lithic tool were the following: size and retouch of the edge and the artifact in general (e.g. bifacial or unifacial work), the base shape and raw material. The specimens were associated in assemblages and sub assemblages, particularly for their temporal relation. As well as this, within the lithic tools, there was a rearrangement based on the shared traits. This is the case of unifacial lanceolate artifacts which compose one class. This class is defined by a general unifaciality of the artifacts and the lanceolate morphology of lateral edges parallels or sub parallels. Other trait presented in a 86.2% of the specimens (N=30) it's a little retouch in the ventral face, generally in the proximal part. Probably this retouch is related to lower the bulb for putting a shaft. Also in several cases, the specimens present apices that could be related with different functions, for example as extractive technologies. The analysis of the variability in the unifacial lanceolate artifacts allow to prove hypothesis about the bias in the

cultural transmission in the population level and along a certain period of time.

These tools came from different sectors of Pastos Grandes and near areas. In special, they were recuperated in La Hoyada, in the component of the Alero Cuevas dated in ca. 4200 BP, in survey recollections and in an area next to Pastos Grandes, San Antonio de los Cobres. Specifically, the site located in San Antonio de los Cobres is called Ramadas (Cardillo 2004, Muscio 2004), with survey and layer's materials dated in ca 5200 BP (Muscio 2004). Therefore, from the chronology available it is possible to estimate a temporal block between 5200-4200 BP for the presence of unifacial lanceolates in the Puna of Salta.

The base shape of these artifacts would correspond to blades. This is supported by the presence in the archaeological record of blades and blade's cores.

Also, I distinguished among tools of high and low cost (Bousman 1993, Muscio 2004), taking into account the energetic inversion in the manufacture. The high cost tools present a high level of work, generally with retouch bifacial extended (e.g. extractive technologies as bifacial points). The low cost tools present a low level of work, specifically the artifacts with marginal or discontinuous retouch, and natural and irregular edges. High and low cost aren't considered fixed entities but they reflect a *continuum* between two extremes with variants in the middle. The base shape also can be seen as of high or low cost. For example, the use of blades would demand more energetic inversion than the common flakes.

In the case of unifacial lanceolate artifacts, the base shape is costly (blades), but the energetic inversion in the work of the edges is generally of a lower cost (retouch marginal and natural edges).

To analyze the importance of the social learning in these artifacts, I have studied the variability in the metric traits. It is expected that while the social learning associated with a strong bias in cultural transmission increases, the variation will be reduced obtaining similar products. This is expected due to the complexity in the preparation of cores and the production of technology of blades.

The principal metric variables used here correspond to the maximum length, maximum width and maximum thickness. Also I considered the length and angle of the edges and specially the index between width and length which gives an idea about of the shape (Cardillo 2004). From this index I established the variation in the lanceolate module toward shapes more or less elongated. If the width-length index increases, it will produce a greater width in relation to the length and vice versa.

The metric variability was analyzed through the variation coefficient (VC), due that is a measure that allows to obtain an absolute value of the total variation of the sample. In addition, I considered the standard residuals to obtain information about the variation of each variable between them and in relation to the average in the regression line.

The width is the measure of less variation, with only 0.14 (14 %). The VC average between length, thickness and width also indicate a low variation (0.18-18 %). The length of the edges has a high correlation with the maximum length of the specimens (R=0.96), probably indicating that the aim would have been to obtain lateral parallel large edges for a multifunctional use. Furthermore, the individuals would have used some specimens with natural edges without retouch. The width-length index also have a tendency toward a low variation coefficient (0.16-16 %). The mean of the width-length index is 0.38 and the standard deviation is 0.06, showing a tendency to obtain modules where the length is more than twice the size of the width.

The standardized residuals (Bettinger and Eerkens 1997) showed that the width is the less variable measure with -1.11 while the length is 0.3 and the thickness is 0.8. It is reminded that the positive high values are the more variable and negative high values are the less variable. In this context, the lithic assemblages of La Hoyada and Alero Cuevas, in Pastos Grandes, show us about the human decision making and about the strategies implementted in the develop of the lithic technology. In particular, the role of the unifacial lanceolate artifacts in the human adaptation to the environment. The comparison between different assemblages will allow to make an approach to population level through its occupational history.

In La Hoyada, it was analyzed a sample of 304 tools, cores and debitage, which were recuperated in a survey context with total absence of pottery. This context might probably be temporally associated to the component of ca. 4200 BP of the Alero Cuevas, owing to the fact of the presence of the unifacial lanceolate artifacts and blade's technology. This does not implicate that all the archaeological record of La Hoyada corresponds to only one temporal block. However, it is important to compare La Hoyada with different archaeological contexts of Pastos Grandes for to establish a approach to the areal archaeological structure. In this sense, the assemblage of La Hoyada allow us to analyze the sameness and differences with the lithic record and the components dated in the Alero Cuevas.

The sample analyzed presents a majority of low cost' s tools (86.58%), almost all of them for processing (not extractive). In the debitage it predominated the flakes but also there are blades and laminar flakes (7.38 %). As well as this, the 17.64% of the cores correspond to blade' s cores. The unifacial lanceolate artifacts represent 12.19 % of the tools.

The layer F2 of the Alero Cuevas dated in ca 4200 BP, have similar traits to La Hoyada. There are also a predominance of low cost's tools (66.66 %). The unifacial lanceolate artifacts represent the 26.6 % of the tools. Moreover, it was recuperated only one core in this layer which corresponds to a blade's extractions. I suggest that the blade's and laminar's cores would be related to a planned strategy with the aim of obtaining the lanceolate morphologies. Therefore the greater energetic inversion would be linked to obtain this base shapes while the retouch of the edges in general is of low cost.

With respect to the use of the raw material in the confection of the tools, in La Hoyada and Alero Cuevas dominates different types of local dark basalt. Therefore, the lithic material of La Hoyada is similar in general traits to the component of ca 4200 BP of Alero Cuevas.

The other component considered correspond to the layer C1 of the Alero Cuevas dated in 2020 BP. This moment correspond to the Agroalfarero Early Period (González 1977), that reflect the increase in the importance of the pastoralism and the wide use of the pottery. A comparative perspective of the two components of Alero Cuevas is important to analyze change or continuity in the local evolutionary process. The layer of ca 2000 BP in the Alero Cuevas is composed by the continuity of a straw's litter. The high density and diversity of specimens indicate that this component represents a redundant use of this site, with an increase in the intensity of the occupation.

The composition of the assemblage within the lithic tools of this layer is clearly different with respect to the component of ca 4200 BP. Specially, there is a considerable increase in the extractive technology which is distinguished for being of high cost, bifacial, and principally with triangular points with peduncle and little lanceolate shapes. This is related to the increase in the use of the obsidian, which is present specially in the high cost tools. The obsidian would come from a source located in Quirón that is about 40 km. far from the site Alero Cuevas.

It is also very important to consider that the processing tools are of low energetic inversion like in the other component.

On the other hand, it is remarkable as well that in the component corresponding to the ca 2000 BP there are an absence of unifacial lanceolate artifacts which have been changed for triangular morphologies, specially in extractive technology. The lanceolate morphologies did not disappear but are reduced in account and size, only present in bifacial extractive technology. Moreover, there is not evidence of laminar or blade technology.

ARCHAEOFAUNAL ANALYSIS

The analyzed archaeofaunal record corresponds to the Alero Cuevas site because is the only one with different components. This allows to characterize the time variation in the consume of faunal resources between the 4200 BP and the 2000 BP. The assemblages were defined in each component: one belonging to the component of ca 4200 BP and the other belonging to the component of ca 2000 BP.

The aim is to characterize the diversity of taxa present in the assemblages and its relative importance. In terms of Grayson and Delpech (1998) it is the NTAXA and the NISP.

In the layer of ca 4200 BP the 67.2 %, it is represented for camelids (NISP), while the 28 % correspond to artiodactyls in general. In the last case it cannot be determined whether there were camelids or not because the specimens were very fragmented. It is reminded that artiodactyla is the order that include to cervids and camelids. However it is possible that it principally corresponds to camelids due to the fact that there were not cervid's specimens recuperated. In this case if the artiodactyls were regarded like camelids, the NISP would be of more of 95 %. The NTAXA is low due that in the family's level only could be determined the presence of camelids and chinchillids (these last only 3.2 %). This could be related to the low diversity of biomass animal of high return. The camelids are the faunal resources of more high return in the Puna and in general present a high availability. The rankings of resources that was elaborated for the region always present the camelids in the first places (López and Medina 2001, Muscio 1999, 2004).

In the C1 layer of the Alero Cuevas, the NISP show too a predominance of the camelids and the artiodactyls, which represents in total more than the 82 %. In addition, there were no cervid's specimens recuperated in this case. The NTAXA is similar to the other assemblage and between the rodents predominated very little specimens that would correspond to species that entered the archaeological record for taphonomical reasons.

However inside the NTAXA it was not considered the representation of different species of camelids. Therefore, I have taken osteometrical measures that allow to show an approach toward the tendency of this variability inside of the family camelidae. This analysis is important to establish hunt and pastoralism strategies, from wild and domesticated species respectively, recuperated in the archaeological record (López 2003).

The results allowed to obtain at least two groups of size clearly differentiated: one next to the standard values of llama (domesticated camelid) and other next to vicuna (wild camelid). The middle values could correspond to guanacos (also wild camelids). In the two components predominate the specimens of vicuña, even in the 2000 BP, although in this time the pastoralism would have been established like a principal economic strategy in the Puna (Yacobaccio et al. 1997). Interestingly, in the assemblage of the component of ca 4200 BP, the measures of one phalanx present higher or similar values to the current llama.

This component could represent a time key in the camelid's domestication. At this moment it is difficult to distinguish whether some specimens actually represent domesticated camelids or camelids during the process of domestication that could have been changing its size. However this evidence is starting line for analyzing the domestication process, although we would have to continue researching in the Middle Holocen to establish its origins.

DISCUSSION AND CONCLUSIONS

A conjunction of complementary and changing strategies related to adaptative mechanisms of decision making which are maintained for cultural transmission would correspond to the local evolutionary process. From an inclusive evolutionary perspective and considering the archaeological indicators (lithic technology and archaeo-faunas) I suggest the existence of an evolutionary process related to the changes produced for the camelid's dome-stication and the adoption of pastoralism as an economic strategy. This process would have been accompanied for a residential mobility reduction like it is showed in the high intensity in the human occupations in the areal archaeological structure during the later Holocene. The adaptative strategies of decision making and cultural transmission could have benefited the increase of fitness along the time. The success of determined strategies that would allow exceed the threshold or minimum require-ment of energy for the adaptation to the environment would remain for selective mechanism, specially natural selection. From the lithic technology and the zooarchaeo-logical bones I suggest different strategies which are maintained along of the analyzed components. In the lithic technology it is distinguished a strategy that points toward to the reduction in the costs of processing tools. Furthermore, I argue that the unifacial lanceolate artifacts were probably related to a diversity of functions and it could have been used like processing tools of faunal resources. The disappearance of this artifacts in the component of ca 2000 BP could be related to the cost to obtain the base shape of blades. Its maintenance was possible for bias in the cultural transmission but with the increase in the costs of handling for the introduction of the pastoralism like central strategy (high production cost) it could have been necessary the reduction in different costs, for example in the low energetic inversion in the processing tools. Therefore it would produce a balance between the costs and benefits for minimize the risk. In addition, the local population could have increased the returns from the hunt of faunal resources of high ranking (wild camelids). For this, they had to put more energy in the confection of extractive tools (e.g. bifacial points) within a general strategy of risk's minimization heading toward the diversification in the consume of faunal resources, but of high ranking.

Therefore, the lithic technology would have an adaptative role related to subsistence. In this sense, low cost strategies in processing tools and high cost in extractive technology would have promoted the diversification in the consume of faunal resources of high ranking through the hunt and pastoralism.

The increase in the intensification of the consume of camelids along the time would indicate that these strategies could be maintained for natural selection.

Regarding the units of analysis used, the specimens were included in assemblages and sub assemblages, in some cases considering the common attributes (e. g. the unifacial lanceolate tools) and in other cases regarding the radiocarbon dates (e. g. archaeofaunas). The components analyzed in Pastos Grandes correspond to different human occupations represented in Alero Cuevas. The information of La Hoyada and other isolated points of the space, in join with Alero Cuevas allow an approach to the areal archaeological structure. The assemblage of the unifacial lanceolate artifacts is adequate for making inferences in the population level including cultural transmission patterns in different times and spaces. It is a temporal block of at least one thousands years (5200 – 4200 BP), represented at this moment for two components located in different areas. In this case I would suggest connections between the groups of Pastos Grandes and San Antonio de los Cobres through cultural continuity observed in unifacial lanceolate artifacts along the time. Therefore, from the units of analysis considered I conclude that it was possible an approach at population level.

To sum up, I would like to point out the theoretical and methodological contribution from a inclusive evolutionnary perspective, relating the adaptative decision making with the environment and the transmission of cultural information.

Acknowledgements

I am grateful to Hernán Muscio, Marcelo Cardillo, Cecilia Mercuri and Hugo Yacobaccio for the comments of this paper. Also I want to thank for the collaboration of Ariadna Svoboda and Federico Restifo in the analysis of archaeofaunal and lithic materials, respectively. Thanks to Mora Castro for help me in the translation to the English language and to Natalia López for the translation to french language. This research it is possible for a CONICET grant. The AMS dates were possible for the NSF grant, in the AMS Arizona Laboratory.

References

ASCHERO, C. (1983) Ensayo para una clasificación morfológica de artefactos líticos. Ms.

ASCHERO, C. and H. YACOBACCIO (1999) 20 años después: Inca Cueva reinterpretado. En *Cuadernos del Instituto Nacional de Antropología y Pensamiento Latinoamericano* 18: 7-18.

BETTINGER, R. and J. EERKENS (1997) *Evolutionary implications of metrical variation in Great basin projectile points*. Rediscovering darwin: Evolutionary theory and archaeological explanation: 177-191

BOONE, J. and E. SMITH (1998). Is it evolution yet? A critique of evolutionary archaeology. *Current Anthropology* 39: 141-173.

BOUSMAN, B. (1993) *Hunter gatherer adaptations, economic risk and tool design*. Lithic Technology 18: 59- 86

CARDILLO, M. (2004). *Arqueología y procesos de transmisión cultural. Una aproximación teórico-metodológica*. Tesis de Licenciatura en ciencias antropológicas, Universidad de Buenos Aires.

GONZÁLEZ, A. (1977) *Arte Precolombino en Argentina*. Ediciones Valero, Buenos Aires

GRAYSON, D. and F. DELPECH (1998). Changing Diet Breadth in the Early Upper Paleolithic of South-western France. *Journal of Archaeological Science* 25, 1119-1129.

LÓPEZ, G. (2003). Pastoreo y caza en el temprano de la Puna de Salta: Datos osteométricos del sitio Matancillas 2. *En Intersecciones en Antropología* 4: 17-27. Olavarría

LÓPEZ, G. (2005). Descripción breve de la investigación arqueológica en Pastos Grandes, Puna de Salta. *En Intersecciones en Antropología 6*: 219-222. Olavarría.

LÓPEZ, G. (MS). Ocupaciones humanas a lo largo del Holoceno en Pastos Grandes, Puna de Salta, Argentina: descripción del sitio arqueológico Alero Cuevas.

LÓPEZ, G. and M. MEDINA (2001) Optimización y riesgo desde las arqueofaunas: su aplicación en el análisis de estrategias de producción de alimentos de la Puna de salta y del Sector Central de Sierras centrales. Presentado en el *XIV Congreso Nacional de Arqueología Argentina*. Rosario.

MENGONI GOÑALONS, G. (1999) *Cazadores de guanacos de la estepa patagónica*. Sociedad Argentina de Antropología. Colección de Tesis Doctorales.

MUSCIO, H (1999) Colonización humana del NOA y variación en el consumo de recursos: la ecología de los cazadores recolectores de la Puna durante la transición Pleistoceno-Holoceno. En *Revista NayA Novedades de Antropología y Arqueología*: 1-41

MUSCIO, H. (2004). *Dinámica poblacional y evolución durante el Período Agroalfarero Temprano en el Valle de San Antonio de los Cobres, Puna de Salta, Argentina*. Tesis doctoral inédita, Facultad de Filosofía y Letras, Universidad de Buenos Aires.

MUSCIO, H. and G. LÓPEZ (MS) Unidades de análisis arqueológicas en el estudio evolutivo de adaptaciones con economías de producción de alimentos: Un examen de las arqueofaunas de la Quebrada de Matancillas (Puna de Salta). En prensa en *Revista Shincal* n° 7, Escuela de Arqueología, Universidad Nacional de Catamarca.

RAMENOFSKY, A. and A. STEFFEN EDS. (1998) *Unit issues in archaeology: measuring time, space and material*. The university of Utah press.

RICHERSON, P. and R. BOYD (1992). Cultural inheritance and evolutionary ecology. En *Evolutionary ecology and human behavior*, ed. por E. Smith y B. Winterhalder: 61-92. Aldine de Gruyter. New York.

SMITH, E.A. (1992). Human Behavioral Ecology I. *Evolutionary Anthropology* 1 (1): 20-25.

SMITH, E.A. and B. WINTERHALDER (1992). Natural Selection and Decision Making: some fundamental Principles. *Evolutionary Ecology and Human Behavior*. Ed. por E.A. Smith y B.C. Winterhalder, pp. 25-60. Aldine de Gruyter, New York.

YACOBACCIO, H. (2001) la domesticación de camélidos en el Noroeste Argentino. *En Historia Prehispánica Argentina Tomo* 1, editado por E. Berberián y A. Nielsen, pp. 7-40. Córdoba, Editorial Brujas.

YACOBACCIO, H.; MADERO, C.; MALMIERCA, M. and M. REIGADAS (1997) Caza, domesticación y pastoreo de camélidos en la Puna Argentina. *Relaciones de la Sociedad Argentina de Antropología* XXII- XXIII: 389-428.

FINDING CONCORDANCE IN DARWINIAN ARCHAEOLOGIES: AND WHY AN UNIFIED EVOLUTIONARY ARCHAEOLOGY IS BOTH IMPOSSIBLE AND UNDESIREABLE

Herbert D.G. MASCHNER & Ben MARLER

Department of Anthropology, Idaho State University, 921 S. 8th Avenue, Stop 8005, Pocatello, ID 83209-8005, USA

Abstract: Here we attempt to illustrate the difference between inclusivity and unification. Inclusivity is understood here to mean a multiplicity of approaches which share a common goal or theme while unification is understood as self promotion by political positioning under a single banner. We argue that while both of these will happen, we will not support unification at the expense of inclusivity. We also discuss how inclusivity is good for science in the exact same way that variation in general is good for evolution in that it allows for more options and more of "reality" to be covered.
Keywords: inclusivity – evolutionary archaeology

Résumé: Nous tentons ici d'illustrer la différence entre la complémentarité et l'unification. La complémentarité est comprise ici comme une multiplicité d'approches qui partagent un but commun ou un thème alors que l'unification est comprise comme l'auto proclamation d'une position politique sous une seule bannière. Nous argumentons qu'alors qu'une des deux positions doit être prise, nous ne soutiendrons pas l'unification au dépens de la complémentarité. On discute également comment la complémentarité est favorable pour la science dans le même sens que la variation en général est favorable à l'évolution dans la mesure qu'elle permet plus d'options et de mieux s'approcher de la "réalité".
Mots-clés: complémentarité – Archéologie évolutive

INTRODUCTION

Much like creationists see Darwinists as part of a homogeneous enclave of common scientific interests to be opposed, many of those in anthropology and archaeology see those of us who employ Darwinian or evolutionary approaches as a single enclave of opposition to interpretive, post-structuralist, and other humanistic approaches. A simple perusal of the papers in this session demonstrates that this could not be further from the truth. Yet rather than use this greater theoretical opposition to better the field, we have substantiated the larger critiques through divisive and polarizing oppositions within Darwinian archaeologies that have been quite successful in enhancing some of our careers, but have done less to advance the use of Darwinism in archaeology and the greater social and historical sciences. It is time that we, as a community of Darwinists, agree to reject further negative discourse and instead, find the parts of the dialogue that actually answer interesting questions about the past while discarding the parts that do nothing but aggrandize particularistic perspectives. This is an approach that Maschner and Reedy-Maschner (2003), thinking of the greater field ranging from Darwin to Marx to Bourdeau to Steward, termed N-Dimensional Anthropology, an inclusive, generalizing approach that Ian Hodder (personal communication to Maschner) had previously criticized and rejected as a mish-mash theory of archaeology. This argument is along the lines of Kristian Kristianson's (2004) recent attempt at finding common ground between Darwinism and agency (which Mithen attempted in 1989) but Kristiansen missed the opportunity by rejecting Stephen Shennan's (2003) inclusive Darwinism as epistemologically inconsistent when in fact, Kristiansen was simply pointing out political differences, not epistemological ones.

Here we firmly reject criticisms of inclusivity as academic self preservation and instead will argue that the reason we have so many Darwinian archaeologies, and by default, so many perspectives in the greater archaeological dialogue, is not because some are right and some are wrong, but rather, because they each tell us something different about the past. It is only through the overt acceptance of multiple lines of investigation that any understanding of the past will ever be successful. Using data from the western Alaska Peninsula and Aleutian Islands, we argue that evolutionary archaeology's emphasis on cladistics and phylogeny, here used on artifacts and house form, leads to the evolutionary ecology of hunting strategies, which leads to the evolutionary psychology of status competition, which leads to kin-selection and corporate group formation, which leads to analyzing cultural complexity with a complex systems approach to house size variation, which in turn leads to a phylogeny of co-evolutionary social change among the Aleut within a region-wide engineered and behaviorally constructed niche. We argue that diversity is good for science and good for knowledge. We also believe that since there is no selective pressure to create a unified Darwinism, it is difficult to envision how it could come to be. But if we could all agree to agree that by using all of these approaches we will have a much stronger interpretation of the archaeological record, then there would be selective pressure for interpretive and humanistic side of the discipline to consider evolutionary approaches. Thus, as our title clearly states and for all the same reasons that a homogenized species is undesirable, we believe a unified Darwinism is undesirable because for some, unification is synonymous to agreeing with them and not someone else – a situation which causes myopic, self-aggrandizing behavior to supersede scientific progress and a perspective soundly rejected in the book *Darwinian Archaeologies* 10 years ago (Maschner 1996). While recognizing that there are theoretical and intellectual inconsistencies between the major sub-approaches, we also recognize that highly disparate approaches such as evolutionary psychology, evolutionary archaeology, behavioral ecology, or dual inheritance provide equally disparate insights into the past. We believe that an inclusive Darwinism, one that moves seamlessly from the individual to the population, one that simultaneously incorporates multiple spatial and temporal scales, and one that sees everything from artifacts, to kin-selection, to mate choice, to optimal foraging to cognitive evolution and complex systems as contributing to our greater understanding of humanity, will be the sort of unified approach that will make such a powerful contribution to archaeology that other theoretical approaches will pale in comparison.

A DARWINIAN DISCOURSE

So let's begin with evolutionary archaeology's use of phylogeny and historical process in archaeology. We do this not because we place any particular emphasis on the idea that selection acts upon artifacts, as some proponents of that perspective do, but because we are convinced by their argument that culture history is critical to our understanding of behavioral change, and that there are indeed evolutionary relationships between things, regardless of whether variation in those things is a product of human intention, random variation, adaptation and optimality, or some combination.

Phylogenetic analysis allows us to model the pattern of evolution. In this sense, as a methodology, it is not necessarily Darwinian and makes no particular claims as to the process of evolution. In fact, as O'Brien and Lyman point out, phylogeny does not require any particular metaphysic (2003), so any metaphysic that emphasizes change over time and supplies a process should combine with our models and the methodology of phylogenetic analysis should not supersede our understanding of the processes of evolution. In other words, any of our existing theoretical formulations will potentially be sufficient for phylogenetic explanation. As such, the issue becomes what scale we want to use, what data we wish to look at, and what assumptions we are capable of making. These conditions will alter which theoretical orientation we use for explanation. This, in turn, focuses our perspective on a particular medium (out of a selection of many possible mediums) that the process of evolution takes place on, for whatever analysis one is doing.

We do not need to appeal to any particular theory in order to conduct phylogenetic analyses. Evolutionary ecology, for example, is well served with the use of phylogenetic modeling, as phylogenetic models can be created to illustrate the transmission of both cultural and genetic traits showing fitness, flow and transmission of both of these through a population. This allows for an analysis to be conducted as to what features, traits, behaviors (or

anything else that is transmissible) are adaptive and, more specifically, which ones are more optimal in which context and relative to which selection pressures.

A long history of research on the Alaska Peninsula in the western Gulf of Alaska has recovered over 10,000 surface depressions, 4300 villages, and 20,000 stone, bone, and ivory tools spanning 6000 years providing a powerful data set for investigating the development of things, and by proxy, human behavior (Maschner 1999a, 1999b, 2004; Maschner and Bentley 2003; Maschner and Reedy-Maschner 2005). The stone end blades, harpoons, and houses along the north Pacific are thus a perfect dataset for constructing phylogenies of material remains and behaviors (work in progress). While a phylogeny of stone tools might be interesting and has shown us interesting preliminary patterns, it is the phylogeny of harpoons that lead us to new levels of analysis. These technologies are a measure of the nature of marine mammal harvesting that provides the foundation of the social worlds of north Pacific peoples. Fixed harpoons without line attachments to fixed harpoons with line attachments to toggling harpoons directly relate to changing hunting strategies. These strategies reflect changing levels of intensification in relation to the foraging efficiency of the people involved. This in turn leads to evolutionary ecology, which is primarily concerned with measuring fitness levels of individuals (sometimes groups of individuals) relative to the behavioral strategies these individuals engage in. This particular point of view rests on the concept of optimization which assumes that individuals will engage their environment in ways (either intention-nally or otherwise) meant to maximize their reproductive success by finding an optimum strategy for engaging their environment (in the most common version, optimal foraging, the scale used to judge this is food procurement as a ratio of energy expenditure to caloric intake) (Bentley *et al.* In Press; Boone 1992; Shennan 2003; Smith and Winterhalder 1992). As is the case with most theoretical generalizations, this has been heavily criticized as not empirically factual but if we keep in mind that this is nothing but a theoretical generalization we can still accept this premise. As Stephen Shennan tells us "It is important to understand that its [evolutionary ecology] advocates are not claiming this as a valid empirical generalization about the state of the world. In fact, it is a framework for generating hypotheses" (Shennan 2003:24). In looking at the north Pacific data, a clear trend in sea mammal hunting on the north Pacific is a trend towards the harvesting of larger and larger mammals has been shown. The earliest sites are dominated by small seals, while fur seals and sea lions increase in importance through time, followed by and increasing reliance on whaling. The evolution of the harpoon technology correlates strongly with the growing importance of larger and more dangerous prey; a pattern easily predicted through evolutionary ecology.

Expanding our use of phylogenies, we created an evolutionary tree of house and hearth form for the entire

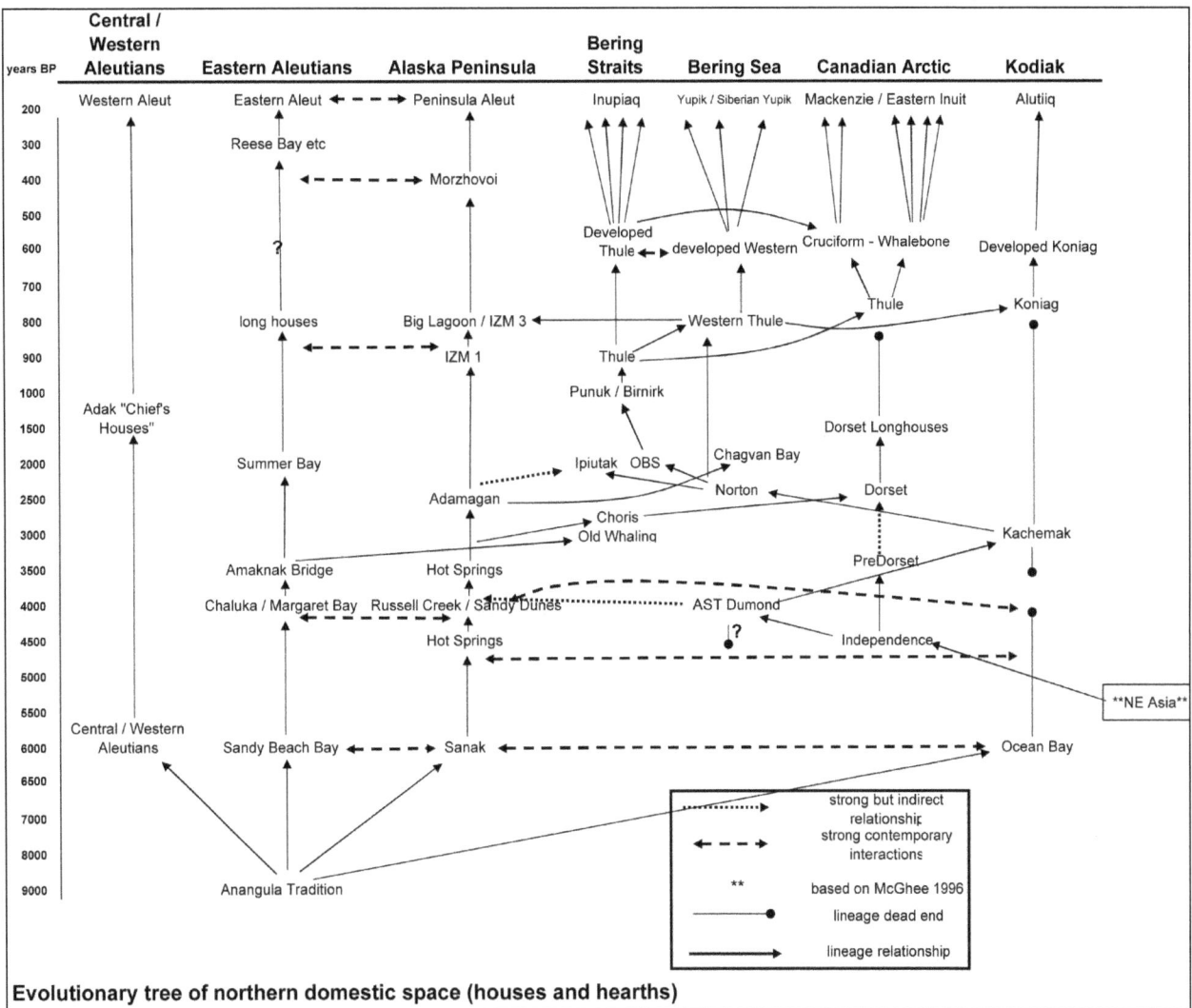

Fig. 8.1. An Evolutionary Tree of Arctic House and Hearth Form

arctic region (Fig. 8.1). A phylogeny of house form reflects changes in household size and organization, and thus changes in the social and political worlds of northern peoples. The small houses of 8500 years ago change little over the next 4000 years, reflecting a long-term pattern of nuclear family organization within a village of similarly sized and organized entities. But after 4500 years ago, house form and size changes through time in response to major shifts in politics, demography and the distribution of resources. By 2500 years ago, we find extended families in centralized villages with evidence of status differentiation in some areas, but small tent camps in others. By 900 years ago in the Aleutian Islands there are massive corporate households organized around entrenched nobility with a class of slaves inhabiting towns of marine foragers while in the eastern arctic this period ushers in 500 years of diversification in house form and organization. It is by thinking historically and phyloge-netically we see the linear relationships between things, in this case households, which forces us to think temporally, or evolutionarily. But in explaining these changes, we turn not in this case to evolutionary ecology, which was

important in understanding the phylogenies of harpoons, but rather to the evolution of corporate group formation explained by kin selection.

Ten years ago John Patton and Maschner (1996) argued that kin-selection might be a powerful means for understanding the rise of large corporate households and social rank on the north Pacific. This was elaborated by Maschner and Hoffman (2003) in their comparative study of households across the entire region and through time. Evolutionary psychology was employed to assess the importance and ultimate foundation for the status competition or aggrandizing behavior necessary for this type of model to be useful (Barrett et al. 2002; Tooby and Cosmides 2005). Once the evolutionary psychology of status competition was understood, then fashionable terms such as costly signaling became irrelevant. Costly signaling is an attempt by evolutionary ecologists to rationalize certain aggrandizing activities and see them as adaptive. But if we look to the greater Darwinian field, we find that the cognitive adaptations that result in these behaviors may have no adaptive significance in more complex societies except for the general pattern that it is simply good to be the king.

Maschner and Alexander Bentley (2003; Bentley and Maschner 2003) took this one step further and identified power laws of household size that, while recognizing that households become generally larger through time, also demonstrated that the presence of large houses as a measure of social power have a 4000 year history. We used Bodley's "the rich get richer" model (1999) to show that there is a wealth disparity in social power beginning in the earliest villages and demonstrated that a complex systems approach to evolutionary change was critical to our explanations of the development of Aleutian households. To understand the complete picture of change on the north pacific, integrating evolutionary ecology, evolutionary archeology, evolutionary psychology, and complex systems becomes the only viable option.

Further, one system of inheritance that we see as becoming more important to archaeology in the near future is that of environmental inheritance. Sometimes called triple-inheritance theory by Riede (2005) or more specifically niche construction as formulated by Odling-Smee *et al* (1999, 2003) Niche construction in archaeology focuses on how humans alter their environment (behavior) by looking at material culture remains as a system of transmission (so no appeal to cognition is necessary). When humans create artifacts they alter their environment and the environment of conspecifics and descendants (cultural and genetic), thus changing many of the selection pressures (again, both cultural and genetic) on all of these and themselves.

For the last few years Maschner has led the National Science Foundation funded Sanak Biocomplexity Project which integrates a suite of social, natural and physical sciences. The basic theme of the project is that the indigenous inhabitants of the north Pacific have been harvesting the region for so long that the "natural" ecology of the north Pacific cannot be understood without reference to the Aleut people as ecosystem engineers. In fact, we have demonstrated that the "natural" behavioral ecology of many sea mammal species is actually a product of adapting to Aleut harvesting intensity. This process is perhaps best understood using Odling-Smee *et al's* concepts of niche construction because it is becoming clearer that the "natural" world harvested by the Aleut peoples of the Aleutian Islands is actually an engineered by-product of that harvesting created over thousands of years.

Thus, through the use of phylogenetic and evolutionary trees, one of the hallmarks of evolutionary archaeology, through the use of foraging theory and kin-selection, both fundamental to evolutionary ecology, by invoking evolutionary psychology to better understand the assumptions associated with the status competition, by using the modeling power of complexity theory, and using niche construction to investigate the role the Aleut played in the engineering of the north Pacific ecosystem and their social worlds, we have demonstrated the usefulness of a number of Darwinian approaches without violating the basic assumptions of any of them.

A DARWINIAN AGENDA

"Blasphemy protects one from the moral majority within, while still insisting on the need for community. Blasphemy is not apostasy. Irony is about contradictions that do not resolve into larger wholes, even dialectically, about the tension of holding incompatible things together because both or all are necessary and true" (Haraway 1991: 149).

In the early 1900's Sigmund Freud was the (relatively) unchallenged pontiff on human psychology. He, as such people tend to do, was losing sway over some of his followers. One of the most influential of these was the famous mystic Carl Gustav Jung. Up to this point these two men had a very close relationship as mentor and protégé. Freud was an aging scholar grooming his student to take over for him when he got too old to continue his work. Jung was a younger, but equally brilliant scholar who would be, in all respects, it seemed at the time, a possible replacement for Freud as the High Priest for this group. Unknown to his teacher, Jung was harboring ideas of his own about what sorts of things were jingling around in the human sub- and unconscious and about the causes of human behavior. Basically, Jung was coming to realize that, contrary to Freud, human psychology was not fully explained by sexual obsessions and neuroses. One of the most famous of those identified by Freud is the Oedipus complex, which was Freud's theory about young males supposed subconscious desire to kill their fathers and to replace them.

Now, this is not a unique situation in the history of ideas or in the history of mentors and students, however, what happens next is worthy of being called a parable. In Denmark, at a dinner Jung chose to unburden himself of aspects of his new theory. This sent Freud into a fit and he fainted in front of the entire assembly. It seems that Freud believed so deeply in his theories that when his son-like protégé challenged him, his Oedipus model kicked into gear and his interpretation of this was that Jung was metaphorically attempting to "kill" him (Jung 1976).

There are two interesting points about this story we would like to emphasize. The first is the fact that epistemological fanaticism mixed with institutionalized political hierarchies and the competition for jobs makes theoretical discussions downright dangerous to one's health. But we also must be aware and concerned about the hypnotic power of one's pet theory or model. We must recognize that all of the different approaches to a Darwinian archaeology share a common foundation and a common logic. They emphasize different aspects of Darwinian thinking which, itself emphasizes particular aspects of the universe, and as such are useful in different situations. In some arenas human intention/agency may be unimportant or overly confusing to analysis, this does not mean or even suggest that intention will be adaptively neutral in all situations (or that intention does not exist) but rather, that it will alter our strategies for generating and testing hypotheses. In some cases optimal behavior may be reached, however this may not always be the case and the concept of optimal may be better understood (in a wonderfully Darwinian manner) as a range. Most if not all of us can probably agree that some aspects of human cognition have evolved over our phylogenetic history and that this will translate into some influence on human behavior. Scientists must keep their options open and realize that Science is, after all, the methods and theories used to find and explain facts, not the facts themselves. Alfred Korzybski has said in his book *Science and Sanity* "The map is not the territory" (Korzybski 1941). Not only is the map not the territory but no map is capable of representing the entire territory. As "the worlds friendly genius" Buckminster Fuller has been kind enough to point out in slogan form "Scenario Universe is non-simulta-neously apprehended" (Fuller 1975).

"Darwinism," "Darwinist," "Darwinian" are not "natural kinds" (neither are behavioral ecology, evolutionary psychology, or evolutionary archaeology) they are terms that refer to the map not the territory. This is not to say that the map and the territory do not influence each other or to go the route of Jean Baudrillard (1994) and discuss a radical dualism between the fully unconnected "Real" and the "Hyper-real" but rather to suggest that some of the "institutional facts," as John Searle (1995) refers to human constructed facts, do not have scientific epistemological intentions behind them. Some lines of our map demark territorial squabbles and not geographic contours "...forcing us (who despise polemics) at last to indulge in a Plenary Session devoted to denunciations ex cathedra, portentous as hell; our faces burn red with rhetoric, spit flies from our lips, neck veins bulge with pulpit fervor. We must at last descend to flying banners with angry slogans (in 1930's type faces)..." (Bey 2003). But here we need a wider view that encompasses all of our approaches while not limiting them or sublimating one or more approach to another equally limiting approach. If we, as Darwinists trying to extend our chosen theoretical orientation out of the solely Biological version of Darwinism know anything, it is that one version of Darwinian thinking does not explain all of Reality. As Gabora states, "Biological evolution offers about as complete an explanation for the existence of dishwashers as the physical constraints and self-organizing properties of matter offers for the existence of giraffes" (Gabora 2001).

We find ourselves in the situation where we must ask ourselves what it is about "Darwinism" which makes us all self associate as "Darwinists?" It is our contention that the unifying force allowing us to be a community is not particular to any of the various groups who identify

themselves as the "true Darwinists." Whatever ideological genealogies (phylogenies) scientists and other academics construct showing them as cladistically closer to whomever is our high-priest is meaningless when compared to what we should be concerned with. What do these models tell us about reality? Is this information useful in a particular "non-simultaneously apprehended" case study? Historiography is useful but it is not science.

The unifying force is the interest in the process of evolution and the acceptance of the fact that at the very least some aspects of it are guided by blind, dumb, unintentional forces with absolutely no foresight, intention or agency what-so-ever. There have been many distillations of evolutionary logic proposed and used. Our current favorite is the abstract information theory approach discussed by Henry Plotkin and others. This allows all current and possible Darwinisms to be linked in an embedded hierarchical fashion with each as a source of diversity and entropy for the others. The logic of evolution is usefully abbreviated into what is called the g-t-r heuristic. That is, any Darwinian system must at very least have a generation phase, a testing phase and then it must regenerate based on the testing phase. The medium neutral terminology and logic allows for this to be applied wherever a range of variation exists.

We do not wish to be understood as claiming that some of the theoretical discussions within Darwinism are not important in all sectors and to all versions of Darwinism. For example the debate over whether or not a replicator is necessary in a system for said system to be Darwinian and if any, what form this replicator must take, whether or not a certain relative amount of fidelity is required, binary and/or particulate versus analogue and/or continuous transmission, memes versus world views versus characters and character states etc. are indeed important. Also, the debate over transmission styles and their effect on rate, tempo and pattern of evolution will affect any Darwinian system. Indeed, to cultural evolution and specifically archaeological analysis the role which artifacts play is of utmost importance. It has been suggested that artifacts are part of the human phenotype, that they are replicators, that they are proxy measures of human behavior, that they are environmental aspects of humans, or that they "extend" the human body (in the case of epistemic artifacts, extend the human nervous system) in both physical and virtual space. It is our contention that any of these are possible *a priori* and that artifacts can be all of these at one time. The question is in what particular circumstances it would be useful to describe a particular set of artifacts in what particular way.

One potentially useful way we might do this is to return to one subset of niche construction that is emphasized by Sterelny (2003, 2005) is that of epistemic artifacts. When an artifact becomes laden with meaning it has the ability to alter not only the physical environment but also the symbolic environment in several ways, including external memory devices where artifacts are used as anchors for schemas (Sterelny 2003, Maschner and Marler in press) allowing for a lessening of cognitive load in many important human tasks by limiting the amount of "new" information humans must deal with while performing vital tasks. As Kim Sterelny points out, many of the evolutionary ecological models of optimal foraging do not take into account these types of energy expenditure and energy saving tactics. If they did, this would be a powerful point of merger between evolutionary psychology and evolutionary ecology. The triple inheritance system allows a foundation where we can build a "unified" Darwinian archaeology by allowing all of the three currently popular approaches (evolutionary archaeology, evolutionary psychology, and evolutionary ecology) to have their own particular aspect of a larger system to describe. The use of all three approaches will aid us in creating a more accurate picture of human prehistory, history, as well as modernity. So, in this sense, we advocate a "unification" of Darwinian approaches to archaeology through the perseverance and maintenance of the diversity of these approaches.

AN INCLUSIVE CONCLUSION

Several Darwinian inheritance systems have been suggested such as genetic, cultural (memetic in some cases), epigenetic, behavioral, ecological/environmental, artifact (both assemblages and characters) and mental "worldviews." We are willing to entertain any of these as a level of analysis and as a potential Darwinian system as long as doing so is capable of telling us something useful about the universe. In the abstract view suggested here one can decouple the process of evolution from any particular medium allowing them to think about evolution using the concepts and terminology of information theory. In this way of thinking a transmission system is any system where information is passed from one generation to the next and a Darwinian system is a transmission system where the system has the ability to store information which is tested versus other similar types of information in some context and differential fitness takes place with the outcome being the baseline for the next generation.

For many, until there is selective pressure for people to agree on what constitutes an appropriate Darwinian archaeology, there is certainly no reason to agree and immense selective pressures not to. That is, since status and prestige are measured by publications and grants, and publications and grants are easiest when one has another publication to argue against, then the argument becomes more important than any explanation of the past that might be put forward using that approach. What archaeology must do is stop publishing the polemic, and only publish the results. The only way one can learn something about the past is to recognize that all of the various forms of Darwinian Archaeologies tell us something interesting about our world. So, is a unified Darwinian Archaeology possible? We do not think one is necessary. But what is absolutely critical to the field is a group of unified Darwinian Archaeologists. A group that finds the past more interesting than the self-aggrandizement gained by arguing either a ridiculous polemic or through trashing their colleagues in the literature. Only then will the panorama of quality Darwinian Archaeologies become relevant to the greater human sciences.

References

BARRETT, LOUISE, ROBIN DUNBAR and JOHN LYCETT (2002) *Human Evolutionary Psychology*. Princeton University Press.

BAUDRILLARD, JEAN (1994) *Simulacra and Simulation*. University of Michigan Press.

BENTLEY, R. ALEXANDER and HERBERT D.G. MASCHNER (2003) Complex Systems and Archaeology: Empirical and Theoretical Applications. Salt Lake: University of Utah Press.

BENTLEY, R. ALEXANDER, CARL LIPO, HERBERT D.G. MASCHNER, and BEN MARLER (In Press). Darwinian Archaeologies. In R. Alexander Bentley, Herbert D.G. Maschner, and Christopher Chippinale, editors. *The Handbook of Archaeological Theory*. Altamira Press.

BEY, HAKIM (2003) *Temporary Autonomous Zones: Poetic Terrorism and Ontological Anarchy*. Autonomedia.

BODLEY, JOHN H. (1999) Socioeconomic Growth, Culture Scale, and Household Well-Being: A Test of the Power-Elite Hypothesis. *Current Anthropology*. 40(5):595-620.

BOONE, JAMES L. (1992) Competition, Conflict and Development of Social Hierarchies. *Evolutionary Ecology and Human Behavior.* Eric Alden Smith and Bruce Winterhlader editors. Aldine De Gruyter. New York, New York.

FULLER, BUCKMINSTER and E.J. APPLEWHITE (1975) *Synergetics: Explorations in the Geometry of Thinking*. Scribner. New York, New York.

GABORA, LIANE (2001) *Cognitive mechanisms underlying the origin and evolution of culture*. Doctoral Thesis. Free University of Brussels.

HARAWAY, DONNA (1991) A Cyborg Manifesto: Science, Technology, and Socialist-Feminism in the Late Twentieth Century. in *Simians, Cyborgs and Women:* The Reinvention of Nature. Routledge. New York, NY.

JUNG, CARL GUSTAV (1976) *The Portable Jung*. ed. Joseph Campbell. Viking Portable Library.

KORZYBSKI, ALFRED (1941) *Science and Sanity: An Introduction to Non-Aristotilian Systems and General Semantics*. 2nd Edition International Non-Aristotelian Publishing Co. Laxeville, Connecticut.

KRISTIANSEN, KRISTIAN (2004) Genes versus agents. A discussion of the widening theoretical gap in archaeology. *Archaeological Dialogues* 11 (2) 77-99.

MASCHNER, HERBERT D.G. (1996) *Darwinian Archaeologies*. Plenum Press: New York.

MASCHNER, HERBERT D.G. (1999a) Prologue to the Prehistory of the Lower Alaska Peninsula. *Arctic Anthropology*. 36(1-2):84-102.

MASCHNER, HERBERT D.G. (1999b) Sedentism, Settlement and Village Organization on the Lower Alaska Peninsula: A Preliminary Assessment. Pp. 56-76. In B. Billman and G. Feinman (editors). *Settlement Pattern Studies in the Americas: Fifty Years since Viru*. Washington: Smithsonian Institution Press.

MASCHNER, HERBERT D.G. (2004) Traditions Past and Present: Allen McCartney and the Izembek Phase of the Western Alaska Peninsula. *Arctic Anthropology*. Vol. 41, No. 2, Pp. 98-111.

MASCHNER, HERBERT D.G. and R. ALEXANDER BENTLEY (2003) The Power Law of Rank and Household on the North Pacific. In *Complex Systems and Archaeology: Empirical and Theoretical Applications*. Pp. 47-60. Edited by R. Alexander Bentley & Herbert D. G. Maschner. University of Utah Press, Salt Lake City.

MASCHNER, HERBERT D.G. and BRIAN W. HOFFMAN (2003) The Development of Large Corporate Households along the North Pacific Rim. *Alaska Journal of Anthropology*. Volume 1(2).

MASCHNER, HERBERT D.G. and BEN MARLER (In press) Evolutionary psychology and archaeological landscapes. in *The Handbook of Landscape Archaeology*. Edited by Bruno David and Julian Thomas. Left Coast Press.

MASCHNER, HERBERT D.G. and JOHN Q. PATTON (1996) Kin selection and the origins of hereditary social inequality: A case study from the Northern Northwest Coast, in H.D.G. Maschner (ed.) *Darwinian Archaeologies*: 89-107. New York: Plenum.

MASCHNER, HERBERT D.G. MASCHNER and KATHERINE L. REEDY-MASCHNER (2005) Aleuts and the Sea. *Archaeology*. March/April pp. 63-70.

MASCHNER, HERBERT D.G. MASCHNER and KATHERINE L. REEDY-MASCHNER (2003) Building an N-Dimensional Anthropology. *Anthropology Newsletter*. 44:9, 4-5.

MITHEN, STEVEN J. (1989) Evolutionary Theory and Post-Processual Archaeology. *Antiquity*, 63, Pp. 483 – 494.

O'BRIEN, MICHAELJ and R. LEE LYMAN (2003) *Cladistics and Archaeology*. University of Utah Press. Salt Lake City, Utah.

ODLING-SMEE, F. JOHN, KEVIN N. LALAND and MARCUS W. FELDMAN (1999) Evolutionary consequences of niche construction and their implications for ecology. *Proceedings of the National Academy of Sciences*. Vol. 96. 10242–10247.

ODLING-SMEE, F. JOHN, KEVIN N. LALAND and MARCUS W. FELDMAN (2003) *Niche Construction*. Princeton University Press. Princeton, New Jersey.

PLOTKIN, HENRY (1995) *Darwin Machines and the Nature of Knowledge*. Harvard University Press. Cambridge, Massachusetts.

RIEDE, FELIX (2005) Darwin versus Bourdieu – celebrity death match or postprocessual myth? A prolegomenon for the reconciliation of agentive-interpretive and ecological-evolutionary archaeology. in L. Grimshaw & F. Coward (eds.) *Investigating Prehistoric Hunter-Gatherer Identities: Case Studies from Palaeolithic and Mesolithic Europe*. BAR (IS) 1411, pp. 45-64.

SEARLE, JOHN R. (1995) *The Construction of Social Reality*. The Free Press. New York, NY.

SHENNAN, STEPHEN (2003) *Genes, Memes and Human History: Darwinian Archaeology and Cultural Evolution*. Thames and Hudson. New York, NY.

SMITH, ERIC A. and BRUCE WINTERHALDER (1992) *Evolutionary Ecology and Human Behavior*. New York: Aldine de Gruyter.

STERELNY, KIM (2003) *Thought in a Hostile World: The Evolution of Human Cognition*. Blackwell Publishing. Malden, MA.

STERELNY, KIM (2005) Externalism, epistemic artefacts and the extended mind. In (R. Schantz, ed) *The Externalist Challenge: New Studies on Cognition and Intentionality*. de Gruyter. New York, New York.

TOOBY, JOHN and LEDA COSMIDES (2005) Conceptual foundations of evolutionary psychology. In *The Handbook of Evolutionary Psychology*. Ed. David M. Buss. John Wiley & Sons, Inc. Hoboken, New Jersey.

THE EXPERIMENTAL SIMULATION OF ARCHAEOLOGICAL PATTERNS: A CONTRIBUTION TO A UNIFIED SCIENCE OF CULTURAL EVOLUTION

Alex MESOUDI

W. Maurice Young Centre for Applied Ethics, University of British Columbia,
227-6356 Agricultural Road, Vancouver, BC, V6T 1Z2, Canada, Tel: 604-827-3519,
Email: mesoudi@interchange.ubc.ca, Website: http://gels.ethics.ubc.ca/Members/amesoudi

Abstract: In the last few years a number of evolutionary archaeologists have argued that certain archaeological patterns can be seen as the result of the cultural transmission of information and artifacts from individual to individual within populations and across successive generations. This has coincided with the development of experimental simulations of small-scale cultural transmission by certain evolutionarily-minded anthropologists and psychologists. Experimental methods such as these can potentially reveal important insights into the large-scale patterns observed in the archaeological record. Experimental simulations offer a number of advantages not available to archaeologists, such as the ability to 're-run' history more than once, the ability to isolate and control single variables, and the generation of complete data-sets. Used in conjunction with archaeological methods and computer simulations, such simulations can be used, for example, to support inferences regarding the precise cultural transmission mechanisms originally responsible for generating different archaeological patterns. This is illustrated with recent experimental simulations of patterns of Great Basin projectile point variation. Finally, cross-disciplinary work such as this is facilitated by the adoption of a unified Darwinian evolutionary approach to human culture.
Keywords: cultural transmission- experimental simulation- unified Darwinian evolutionary

Résumé: Ces dernières années, un certain nombre d'archéologues évolutionnistes ont avancé que certains schémas archéologiques peuvent être perçus comme le résultas d'une transmission culturelle d'information et d'objets d'individus en individus au sein des populations de génération en génération. Ceci coïncide avec le développement de simulations à petite échelle de la transmission culturelle par certains anthropologues et psychologues évolutionnistes. De telles méthodes expérimentales peuvent potentiellement mettre à jour des questions importantes au sein des schémas à grande échelle observés dans le registre archéologique. Les simulations expérimentales offrent nombre d'avantages non accessibles aux archéologues, telles que la possibilité de 'repasser' l'histoire plus d'une fois, la possibilité d'isoler et contrôler des variables uniques, et la production d'une base de données complète. Utilisées conjointement avec les méthodes archéologiques et les simulations informatiques, de telles simulations peuvent être utilisées, par exemple, pour appuyer des inférences quant aux mécanismes précis de la transmission culturelle originellement responsables de la production de différents schémas archéologiques. Ceci est illustré avec les récentes simulations expérimentales des schémas de variation des pointes de projectile du Grand Bassin. Finalement, un travail interdisciplinaire tel que celui-ci est facilité par l'adoption d'une approche évolutionniste Darwinienne unifiée pour aborder la culture humaine.
Mots clés: transmission culturelle – simulation experimentale – évolutionniste Darwinienne unifiée

INTRODUCTION

In recent years there has been a growing movement within archaeology that seeks to use the theoretical and methodological tenets of Darwinian evolutionary theory to address archaeological problems (Dunnell, 1980; Lipo, Madsen, Dunnell, & Hunt, 1997; Lipo, O'Brien, Collard, & Shennan, 2006; Lyman & O'Brien, 1998; Neiman, 1995; O'Brien, 1996; O'Brien & Lyman, 2000, 2002, 2003; Shennan, 2002; Shennan & Wilkinson, 2001). In a wider context, this movement can be seen as a consequence of treating human culture in general as an evolutionary system, exhibiting the characteristics of variation, selection and inheritance that constitute Darwinian evolution (Boyd & Richerson, 1985; Cavalli-Sforza & Feldman, 1981; Mesoudi, Whiten, & Laland, 2004; Richerson & Boyd, 2005). Within this wider context, archaeology can be seen as the cultural equivalent of paleobiology: just as paleobiologists study past biological evolution using evidence from the fossil record, evolutionary archaeologists study past cultural evolution using evidence from the artifactual record (Mesoudi, Whiten, & Laland, 2006; O'Brien & Lyman,

2000). This parallel can be seen in Figure 9.1, which shows the place of archaeology in the larger evolutionary framework for the study of culture outlined by Mesoudi *et al.* (2006).

Adopting an evolutionary approach to archaeology brings with it a number of related advantages. Methods developed by biologists can be adapted for use with the archaeological record, most prominently cladistic or phylogenetic analyses (Collard, Shennan, & Tehrani, 2006; Lipo *et al.*, 2006; O'Brien, Darwent, & Lyman, 2001; O'Brien & Lyman, 2003) that are used to determine whether similar artifacts share features because of shared ancestry or because they constitute independent innovations in unrelated lineages. For example, O'Brien *et al.* (2001; 2002) have used phylogenetic methods to analyze Paleoindian-period projectile points from the southeastern United States, providing a more methodologically and conceptually robust understanding of the past spread of projectile-point technology in that region than is obtained using alternative methods that do not distinguish between analogous and homologous features, such as phenetics or informal verbal arguments. Tehrani and Collard (2002),

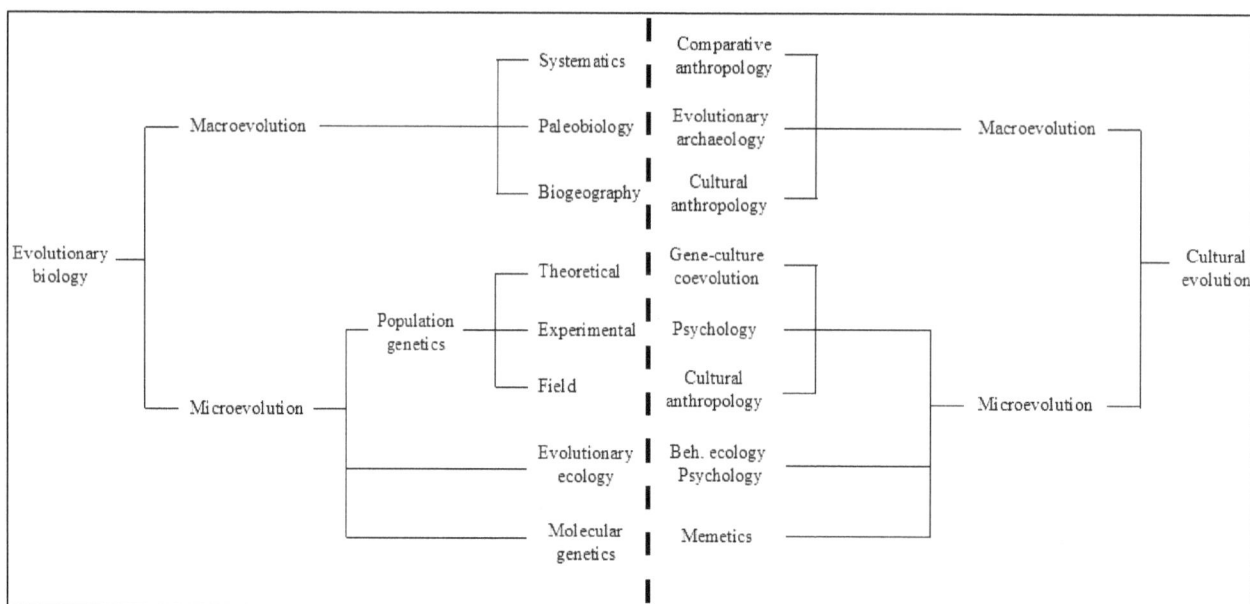

Fig. 9.1. The structure of a unified science of cultural evolution (right hand side), as mapped onto the structure of evolutionary biology (left hand side). Evolutionary archaeology can be seen as the cultural parallel of paleobiology. Adapted from Mesoudi *et al.* (2006)

meanwhile, present a phylogenetic analysis of the patterned designs on Turkmen textiles, finding an earlier period dominated by 'phylogenesis', involving the branching of separate lineages of textile designs, followed by a subsequent period of 'ethnogenesis', involving crossing between lineages. Researchers in this field are also aware that phylogenetic methods cannot be unthinkingly imported from biology; there are important differences between biological and cultural change that necessitate the development of tools specifically designed with cultural data-sets in mind (Borgerhoff Mulder, McElreath, & Schroeder, 2006; Borgerhoff Mulder, Nunn, & Towner, 2006). Another tool that has been borrowed from biology is the neutral drift model (Crow & Kimura, 1970), which has been used by Neiman (1995), Shennan and Wilkinson (2001) and Bentley and Shennan (2003) to identify archaeological change that is consistent with random copying (Neiman, 1995) or that deviates from such an assumption (Shennan & Wilkinson, 2001). These methods have been discussed in detail in the above references, and elsewhere in this volume. Here, I would like to highlight another tool that can be used to explain patterns and trends in the archaeological record – experimental simulations of cultural transmission.

CULTURAL TRANSMISSION AND THE EVOLUTIONARY SYNTHESIS

An evolutionary approach to archaeology naturally leads to a focus on cultural transmission. Transmission, or inheritance, is fundamental to any theory of evolution, be it biological or cultural: "Any variation which is not inherited is unimportant for us" (Darwin, 1859, p. 75). As O'Brien and Lyman (2000) point out, in archaeology this implies that lineages of similar artifacts must be causally linked by cultural transmission (what they call 'heritable continuity'). O'Brien and Lyman (2000) and Lipo et al. (1997) have consequently sought to reintroduce the method of seriation to identify such lineages of artifacts, wherein artifacts are ordered on the basis of their occurrence, frequency, or similarity of appearance. We can go further than this, however, and argue that beyond simply allowing cultural evolution to occur, the details of cultural transmission at the individual level – who copies what, from whom and when – may be responsible for certain long-term, large-scale patterns of change observed in the archaeological record. The fact that the details of information transmission at the individual level can have significant effects at the population-level has been recognized by evolutionary biologists since the evolutionary synthesis of the 1930s and 1940s (Huxley, 1942; Simpson, 1944), and is largely responsible for the unification and subsequent success of the biological sciences during the mid- and latter 20th century (Mayr, 1982; Mayr & Provine, 1980). Similar Darwinian models of cultural change (Boyd & Richerson, 1985; Cavalli-Sforza & Feldman, 1981) send the same message, with the mode of transmission (e.g. vertical, oblique or horizontal: Cavalli-Sforza & Feldman, 1981) and biases in transmission (e.g. conformist or indirectly biased transmission: Boyd & Richerson, 1985) predicted to have significant and identifiable long-term population-level effects. Given that archaeologists are primarily interested in population-level trends and patterns over extended periods of time and across multiple generations, it follows that some of these patterns might be explained by appealing to the individual-level details of transmission (as well as selection, mutation, drift, and other processes, although we focus here on transmission).

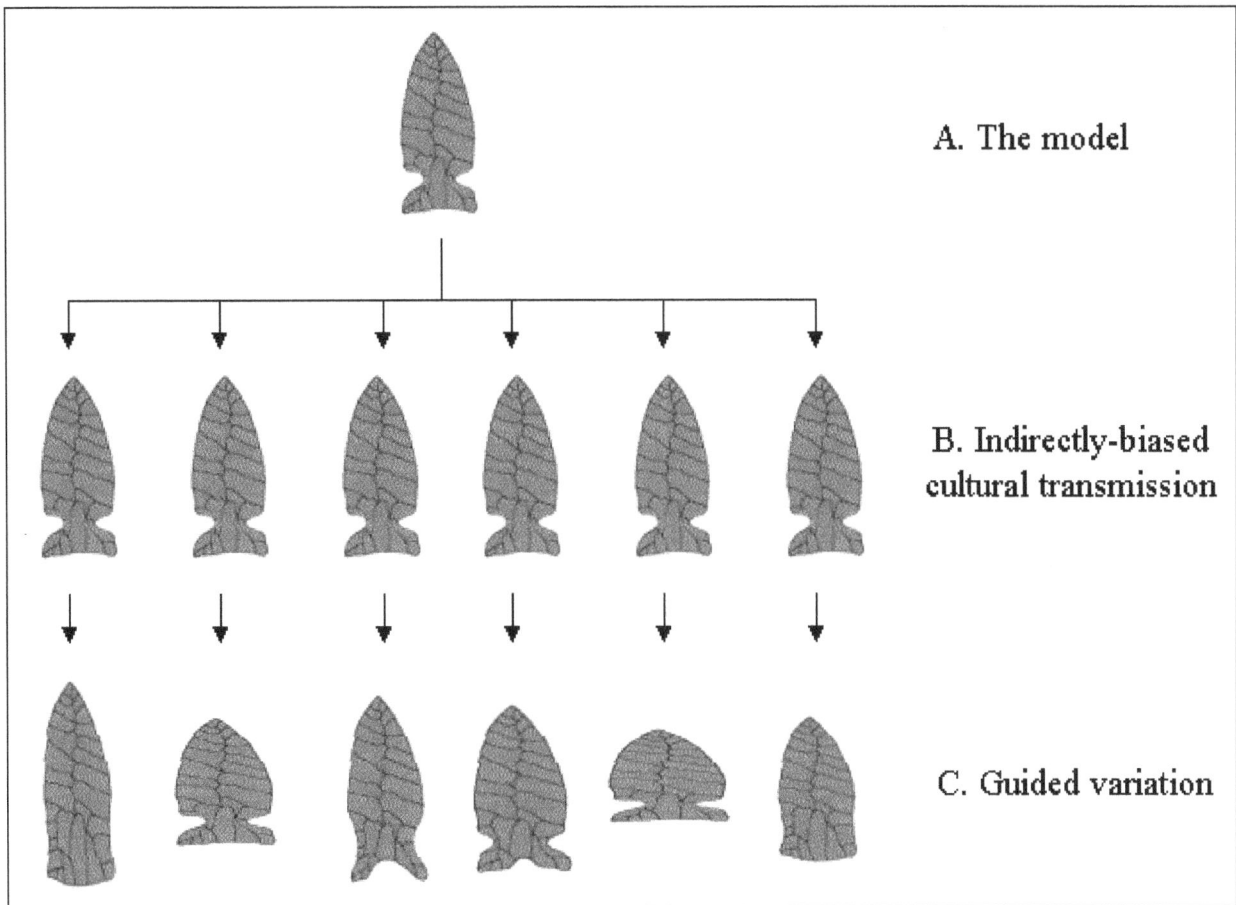

A. The model

B. Indirectly-biased cultural transmission

C. Guided variation

Fig. 9.2. A schematic representation of Bettinger and Eerkens' (1999) hypothesized explanation for differences in prehistoric projectile point variation in the Great Basin. The point design of a successful hunter (A) is copied via indirectly biased cultural transmission (B), resulting in highly correlated point attributes and low variation. In guided variation (C), individuals engage in independent trial and error learning, resulting in increased diversity in point designs and low attribute correlations. In Bettinger and Eerkens' (1999) hypothesis, the stages B and C correspond to points originating from central Nevada and eastern California respectively. Mesoudi and O'Brien (submitted) simulated the learning rules in B and C (indirect bias and guided variation) and confirmed the resulting patterns of variation (low variation in B and high variation in C)

As an example of the use of individual-level transmission biases to explain population-level archaeological patterns, Bettinger and Eerkens (1999) argued that different population-level patterns of variation in Great Basin projectile points can be seen as the result of different means by which the point technology was originally transmitted. Points from eastern California were found to have attributes (e.g. length, width or weight) that were largely uncorrelated with each other, while points from central Nevada had uniform designs with strong correlations between their attributes. Bettinger and Eerkens (1999) argued that the diversity in the Californian sample was generated because point designs in this region origin-nally spread due to 'guided variation' (Boyd & Richerson, 1985), where the point design is copied from a model and is then subject to modification due to individual learning. This modification consequently breaks down any similarities between inherited point designs. Points from Nevada, on the other hand, were argued to have originally spread due to 'indirect bias' (Boyd & Richerson, 1985),

where point designs are copied wholesale from a single successful hunter with no further modification. This hypothesis is illustrated in Figure 9.2.

Although studies such as Bettinger and Eerkens (1999) and modeling work such as Cavalli-Sforza and Feldman (1981) and Boyd and Richerson (1985) explicitly emphasize the role of individual-level transmission modes and biases in generating population-level patterns and trends, similar assumptions underlie much other work in evolutionary archaeology. For example, implicit in the aforementioned studies that have applied the neutral drift model from population genetics to account for specific archaeological trends (Bentley & Shennan, 2003; Neiman, 1995; Shennan & Wilkinson, 2001) is the assumption that at the individual level people are exhibiting a specific transmission rule – copy at random – or some combina-tion of transmission rules that, combined, are indisting-uishable from random copying. Similarly, a major issue in the use of cultural phylogenies is the relative influence of vertical and horizontal transmission (Borgerhoff Mulder, Nunn *et al.*, 2006), i.e. whether people learned from their parents or from unrelated members of their own generation, and whether this transmission occurred between individuals or between groups (Borgerhoff Mulder, McElreath *et al.*, 2006).

EXPERIMENTAL SIMULATIONS OF CULTURAL TRANSMISSION

It should be apparent that in discussing the role of individual-level transmission biases and decision making processes – when, what and who to copy – we are entering the realm of social psychology. The central message of this paper is that psychological experiments have the potential to make a significant contribution to evolutionary archaeology by simulating under controlled conditions the transmission biases and decision processes that are potentially responsible for specific archaeological patterns (see also Mesoudi, in press). Indeed, this cross-disciplinary sharing of concepts and methods constitutes one of the major benefits of adopting an evolutionary approach to human culture (Mesoudi, Whiten, & Laland, 2006). Just as controlled experimental simulations of biological evolution have been used to explain population level patterns in the paleobiological record, such as punctuated equilibria (Lenski & Travisano, 1994), long-term adaptation in rugged fitness landscapes (Colegrave & Buckling, 2005; Elena & Lenski, 2003) and the evolution of sexual reproduction (Colegrave, 2002; Rice, 2002), psychological experiments simulating aspects of cultural transmission can help to explain patterns in the archaeological record (Mesoudi, in press; Mesoudi & O'Brien, submitted). Although experiments have often been used in archaeology in the past to explore the physical characteristics of artifacts (e.g. Cheshier & Kelly, 2006; Flenniken & Raymond, 1986; Knecht, 1997; Mesoudi & O'Brien, submitted; Odell & Cowan, 1986; Schiffer & Skibo, 1987), they have not (to my knowledge) been used to explore the effect of cultural transmission on past cultural change.

Surprisingly, experimental simulations of cultural transmission have been somewhat rare in social psychology, with only a handful of studies carried out during the 20th century (Allport & Postman, 1947; Bartlett, 1932; Insko *et al.*, 1983; Insko *et al.*, 1980; Jacobs & Campbell, 1961). Recent years, however, have seen increased interest in experimental simulations of cultural transmission (Baum, Richerson, Efferson, & Paciotti, 2004; Kameda & Nakanishi, 2002, 2003; McElreath *et al.*, 2005; Mesoudi & Whiten, 2004; Mesoudi, Whiten, & Dunbar, 2006), and point the way to how experiments can be used to study cultural transmission in an archaeological context. In some of these studies, groups of participants are asked to solve some problem or perform some task either through individual learning or by copying a solution from one or more other group members. The experimenter can then determine under what conditions people engage in

individual learning versus cultural transmission/social learning, and examine the adaptive consequences of different cultural transmission biases. For example, Kameda and Nakanishi (2002) had groups of six participants play a simple computer game in which they had to guess in which of two locations a rabbit could be found. Participants had the opportunity to engage in either individual learning at a cost or social learning at no cost. During social learning, the past choices of three randomly selected group members could be seen. As predicted by mathematical simulations, some group members tended to be individual learners ('information producers') and others became social learners ('information scroungers'), with these two sub-groups coexisting. Increasing the cost of individual learning increased the frequency of social learning. In a similar study, McElreath *et al.* (2005) had groups of participants select one of two crops to plant, with the aim of finding the crop that gave the higher payoff despite inaccurate environmental feedback. Participants could (i) engage in trial and error individual learning, (ii) view the past choice of a randomly selected fellow group member (allowing simple cultural transmission), or (iii) view the past choices of all group members (allowing conformist cultural transmission). The results of this study were more equivocal, with most participants not engaging in any form of cultural transmission. Of those who did, conformity was only employed when the environment fluctuated, despite being the optimal strategy under all conditions.

Experiments such as these regarding the relative frequency and efficacy of social versus individual learning and the relative efficacy of different cultural transmission rules, while not tied to specific archaeological data, are obviously relevant to the kind of hypotheses put forward by Bettinger and Eerkens (1999), which concern the same issues in populations of prehistoric hunters. In an initial experimental simulation of archaeological transmission dynamics, Mesoudi and O'Brien (submitted) simulated the different transmission biases argued by Bettinger and Eerkens (1999) to have generated the different attribute correlations in the Great Basin. Participants were asked to design a series of 'virtual projectile points' via a simple computer program and test the performance of their designs in a 'virtual hunting environment'. Different phases of the experiment simulated different transmission biases, giving the participants (i) the opportunity to copy the point design of previous participants given information about those previous participants' relative success in the task (allowing indirectly-biased oblique cultural transmission), then (ii) permitting them to experiment individually with their point designs (simulating guided variation), and finally (iii) letting them copy the designs of other members of their group (allowing indirectly biased horizontal cultural transmission). Consistent with Bettinger and Eerkens' (1999) hypothesis regarding the spread of Great Basin projectile technology, point attributes were more strongly correlated following indirectly biased cultural transmission (both oblique and horizontal) than following individual learning, matching the data-sets from central Nevada and eastern California respectively. This study illustrates the potential value of experimental methods in explaining archaeological change, such as the possibility of replication, the manipulation of variables, access to complete, uninterrupted data, and the comparison of fitness at different points in a lineage (Mesoudi, in press). For example, Mesoudi and O'Brien (submitted) found that increasing the cost of modification (i.e. individual learning) increased the benefit of engaging in indirectly biased horizontal cultural transmission. This provides a potential explanation for the differences between the two archaeological sites, specifically that some aspect of the prehistoric central Nevada environment made experimentation with projectile technology more difficult than in prehistoric eastern California, a hypothesis that could be tested with further archaeological study. Mesoudi and O'Brien (submitted) also found that diversity can be generated and maintained by assuming that point design fitness is determined by a multimodal fitness landscape, where more than one design is locally optimal (i.e. small changes in each design reduces its fitness, discouraging individuals from moving to the single design with the highest fitness). This is potentially consistent with experiments on point performance characteristics (e.g. Cheshier & Kelly, 2006) which find tradeoffs between competing demands, although again further archaeological study is required.

Future studies may well find that alternative assumptions have similar effects, or that some factor missing from our experiment significantly changes the findings. We do not claim in Mesoudi and O'Brien (submitted) to have definitively resolved any issues regarding the spread of Great Basin projectile point technology via our particular experimental simulations, we merely hope to demonstrate that experiments in general provide a means of testing, under controlled, replicable conditions, hypotheses that are extremely difficult or impossible to test using archaeological data alone, such as the underlying fitness functions that drove changes in prehistoric point designs (if indeed point designs were functional, a question which could also be investigated experimentally). The archaeological literature contains a wealth of archaeological patterns that are likely amenable to experimental simulation. Experimental simulations can be used in conjunction with archaeological and ethnographic studies, as well as computer and mathematical simulations, to provide a more complete explanation of past cultural change than any one of these methods alone. This plurality of methods is facilitated by adopting an evolutionary approach to archaeology, and placing archaeology within a larger science of cultural evolution.

Acknowledgements

I would like to thank Michael O'Brien for collaborating on the projectile point experiments, and Hernán Muscio and Gabriel López for organizing the UISPP workshop where this paper was presented.

References

ALLPORT, G.W., & POSTMAN, L. (1947). *The psychology of rumor*. Oxford: Henry Holt.

BARTLETT, F.C. (1932). *Remembering*. Oxford: Macmillan.

BAUM, W.M., RICHERSON, P.J., EFFERSON, C.M., & PACIOTTI, B.M. (2004). Cultural evolution in laboratory microsocieties including traditions of rule giving and rule following. *Evolution and Human Behavior, 25*, 305-326.

BENTLEY, R.A., & SHENNAN, S.J. (2003). Cultural transmission and stochastic network growth. *American Antiquity, 68*, 459-485.

BETTINGER, R.L., & EERKENS, J. (1999). Point typologies, cultural transmission, and the spread of bow-and-arrow technology in the prehistoric Great Basin. *American Antiquity, 64*, 231-242.

BORGERHOFF MULDER, M., MCELREATH, R., & SCHROEDER, K.B. (2006). Analogies are powerful and dangerous things. *Behavioral and Brain Sciences, 29*, 350-351.

BORGERHOFF MULDER, M., NUNN, C.L., & TOWNER, M.C. (2006). Cultural macroevolution and the transmission of traits. *Evolutionary Anthropology, 15*, 52-64.

BOYD, R., & RICHERSON, P.J. (1985). *Culture and the evolutionary process*. Chicago: University of Chicago Press.

CAVALLI-SFORZA, L.L., & FELDMAN, M.W. (1981). *Cultural transmission and evolution*. Princeton: Princeton University Press.

CHESHIER, J., & KELLY, R.L. (2006). Projectile point shape and durability: The effect of thickness:Length. *American Antiquity, 71*, 353-363.

COLEGRAVE, N. (2002). Sex releases the speed limit on evolution. *Nature, 420*, 664-666.

COLEGRAVE, N., & BUCKLING, A. (2005). Microbial experiments on adaptive landscapes. *Bioessays, 27*, 1167-1173.

COLLARD, M., SHENNAN, S., & TEHRANI, J.J. (2006). Branching, blending, and the evolution of cultural similarities and differences among human populations. *Evolution and Human Behavior, 27*, 169-184.

CROW, J.F., & KIMURA, M. (1970). *An introduction to population genetics theory*. New York: Harper & Row.

DARWIN, C. (1859). *The origin of species*. London: Penguin, 1968.

DUNNELL, R.C. (1980). Evolutionary theory and archaeology. *Advances in Archaeological Method and Theory, 3*, 35-99.

ELENA, S.F., & LENSKI, R.E. (2003). Evolution experiments with microorganisms: The dynamics and genetic bases of adaptation. *Nature Reviews Genetics, 4*, 457-469.

FLENNIKEN, J.J., & RAYMOND, A.W. (1986). Morphological projectile point typology: Replication experimentation and technological analysis. *American Antiquity, 51*, 603-614.

HUXLEY, J.S. (1942). *Evolution, the modern synthesis.* London: Allen & Unwin.

INSKO, C.A., GILMORE, R., DRENAN, S., LIPSITZ, A., MOEHLE, D., & THIBAUT, J.W. (1983). Trade versus expropriation in open groups: A comparison of two types of social power. *Journal of Personality and Social Psychology, 44*, 977-999.

INSKO, C.A., THIBAUT, J.W., MOEHLE, D., WILSON, M., DIAMOND, W.D., GILMORE, R., *et al.* (1980). Social evolution and the emergence of leadership. *Journal of Personality and Social Psychology, 39*, 431-448.

JACOBS, R.C., & CAMPBELL, D.T. (1961). The perpetuation of an arbitrary tradition through several generations of a laboratory microculture. *Journal of Abnormal and Social Psychology, 62*, 649-658.

KAMEDA, T., & NAKANISHI, D. (2002). Cost-benefit analysis of social/cultural learning in a nonstationary uncertain environment: An evolutionary simulation and an experiment with human subjects. *Evolution and Human Behavior, 23*, 373-393.

KAMEDA, T., & NAKANISHI, D. (2003). Does social/cultural learning increase human adaptability? Rogers' question revisited. *Evolution and Human Behavior, 24*, 242-260.

KNECHT, H. (Ed.). (1997). *Projectile technology.* New York: Plenum.

LENSKI, R.E., & TRAVISANO, M. (1994). Dynamics of adaptation and diversification – a 10.000- generation experiment with bacterial-populations. *Proceedings of the National Academy of Sciences of the United States of America, 91*, 6808-6814.

LIPO, C.P., MADSEN, M.E., DUNNELL, R.C., & HUNT, T. (1997). Population structure, cultural transmission and frequency seriation. *Journal of Anthropological Archaeology, 16*, 301-333.

LIPO, C.P., O'BRIEN, M.J., COLLARD, M., & SHENNAN, S. (Eds.). (2006). *Mapping our ancestors: Phylogenetic approaches in anthropology and prehistory.* New York: Aldine.

LYMAN, R.L., & O'BRIEN, M.J. (1998). The goals of evolutionary archaeology: History and explanation. *Current Anthropology, 39*, 615-652.

MAYR, E. (1982). *The growth of biological thought.* Cambridge, MA: Harvard University Press.

MAYR, E., & PROVINE, W. (Eds.). (1980). *The evolutionary synthesis.* Cambridge, MA: Harvard University Press.

MCELREATH, R., LUBELL, M., RICHERSON, P.J., WARING, T.M., BAUM, W., EDSTEN, E., *et al.* (2005). Applying evolutionary models to the laboratory study of social learning. *Evolution and Human Behavior, 26*, 483-508.

MESOUDI, A. (in press). The experimental study of cultural transmission and its potential for explaining archaeological data. In M.J. O'Brien (Ed.), *Cultural transmission and archaeology: Issues and case studies.* Washington, D.C.: Society for American Archaeology Press.

MESOUDI, A., & O'BRIEN, M.J. (submitted). The cultural transmission of Great Basin projectile point technology: Experimental and computer simulations. *American Antiquity*

MESOUDI, A., & WHITEN, A. (2004). The hierarchical transformation of event knowledge in human cultural transmission. *Journal of Cognition and Culture, 4*, 1-24.

MESOUDI, A., WHITEN, A., & DUNBAR, R.I.M. (2006). A bias for social information in human cultural transmission. *British Journal of Psychology, 97*, 405-423.

MESOUDI, A., WHITEN, A., & LALAND, K.N. (2004). Is human cultural evolution Darwinian? Evidence reviewed from the perspective of *The Origin of Species. Evolution, 58*, 1-11.

MESOUDI, A., WHITEN, A., & LALAND, K.N. (2006). Towards a unified science of cultural evolution. *Behavioral and Brain Sciences, 29*, 329-383.

NEIMAN, F.D. (1995). Stylistic variation in evolutionary perspective – inferences from decorative diversity and interassemblage distance in Illinois woodland ceramic assemblages. *American Antiquity, 60*, 7-36.

O'BRIEN, M.J. (Ed.). (1996). *Evolutionary archaeology: Theory and application.* Salt Lake City: University of Utah Press.

O'BRIEN, M.J., DARWENT, J., & LYMAN, R.L. (2001). Cladistics is useful for reconstructing archaeological phylogenies: Palaeoindian points from the southeastern United States. *Journal of Archaeological Science, 28*, 1115-1136.

O'BRIEN, M.J., & LYMAN, R.L. (2000). *Applying evolutionary archaeology.* New York: Kluwer Academic.

O'BRIEN, M.J., & LYMAN, R.L. (2002). Evolutionary archeology: Current status and future prospects. *Evolutionary Anthropology, 11*, 26-36.

O'BRIEN, M.J., & LYMAN, R.L. (2003). *Cladistics and archaeology.* Salt Lake City: University of Utah Press.

O'BRIEN, M.J., LYMAN, R.L., SAAB, Y., SAAB, E., DARWENT, J., & GLOVER, D.S. (2002). Two issues in archaeological phylogenetics: Taxon construction and outgroup selection. *Journal of Theoretical Biology, 215*, 133-150.

ODELL, G.H., & COWAN, F. (1986). Experiments with spears and arrows on animal targets. *Journal of Field Archaeology, 13*, 195-212.

RICE, W.R. (2002). Experimental tests of the adaptive significance of sexual recombination. *Nature Reviews Genetics, 3*, 241-251.

RICHERSON, P.J., & BOYD, R. (2005). *Not by genes alone: How culture transformed human evolution.* Chicago: University of Chicago Press.

SCHIFFER, M.B., & SKIBO, J.M. (1987). Theory and experiment in the study of technological change. *Current Anthropology, 28*, 595-622.

SHENNAN, S.J. (2002). *Genes, memes and human history.* London: Thames and Hudson.

SHENNAN, S.J., & WILKINSON, J.R. (2001). Ceramic style change and neutral evolution: A case study from neolithic Europe. *American Antiquity, 66*, 577-593.

SIMPSON, G.G. (1944). *Tempo and mode in evolution.* New York: Columbia University Press.

TEHRANI, J.J., & COLLARD, M. (2002). Investigating cultural evolution through biological phylogenetic analyses of Turkmen textiles. *Journal of Anthropological Archaeology, 21*, 443-463.

A SYNTHETIC DARWINIAN PARADIGM IN EVOLUTIONARY ARCHAEOLOGY IS POSSIBLE AND CONVENIENT

Hernán Juan MUSCIO

CONICET, Instituto de Arqueología, Universidad de Buenos Aires, 25 de Mayo 217 (10020) CABA, ARG
E-mail hmuscio@fibertel.com.ar

Abstract: *A synthetic Darwinian framework, with the potential to link the short term microevolutionary mechanisms to the macroevolutionary patterns documented in the archaeological record, results after discarding a misconceived single-level selectionism. Promoting an expanded Darwinism, adaptationist and populational thinking is the best strategy to modelling the selective environments where humans, artifacts and behaviours evolve thanks to nested selective forces acting at the multiple focal levels of a genealogical hierarchy. After focusing on some of the elements of a multilevel Darwinian framework for evolutionary archaeology, I exemplify in this paper its utility by discussing the evolution of the earliest ceramics of Northwestern Argentina.*
Keywords: synthetic Darwinian framework- selectionism- multilevel Darwinian

Résumé: *Une approche Darwinienne synthétique, avec le potentiel de lier les mécanismes microévolutifs à court terme aux modèles macroévolutifs documentés par le registre archéologique, est possible après avoir rejeté une sélection à niveau simple mal comprise. Promouvant un Darwinisme étendu, penser en termes d'adaptation et de population est la meilleure stratégie pour modeler un environnement sélectif où les humains, les objets et les comportements s'élaborent grâce aux forces sélectives agissant à des niveaux multiples d'une hiérarchie généalogique. Après s'être focalisé sur quelques éléments d'une approche Darwinienne multiniveaux pour une archéologie évolutive, j'illustrerai dans cet article son utilité en discutant l'évolution des céramiques anciennes du Nord-Ouest Argentin.*
Mots clés: approche Darwinienne synthétique – sélection – approche Darwinienne multiniveaux

EVOLUTIONARY ARCHAEOLOGY NEEDS A SYNTHETIC DARWINIAN FRAMEWORK

Darwin's theory of natural selection produced the unification and the success of the biological sciences. The "Modern Synthesis" fostered the proliferation of a diversity of disciplines to account for the many different evolutionary phenomena of the natural world. Of mayor importance was the realization that macro evolutionary patterns result from microevolutionary mechanisms (Foley 1992), and that Darwinian selection might also operates on several evolutionary entities (Gould 2002).

As Mesoudi *et al* (2006) convincingly argue given that culture exhibits key Darwinian evolutionary properties the structure of a science of cultural evolution should share fundamental features with the structure of biological evolution. Nevertheless, a unified Darwinism did not emerge in anthropology or archaeology. Since the application of the scientific theory of evolution in archaeology is a relatively recent enterprise, implicating different seleccionist frameworks and even national traditions, the discipline still lacks such a synthesizing framework. Hence, a broad spectrum Darwinian interpretation prevails under the label "Darwinian Archaeologies" (Maschner 1996).

Archaeology once was closer to converge to a unified theory of cultural evolution. Under the Culture History paradigm of the Americas, with its focus on artifacts change along space and time, archaeology developed a machinery of methods to document cultural homologous similarities (Lyman *et al* 1997). But the essentialist conception of the evolutionary processes, along with an ethnographic description of the archaeological record was the fatal mistakes of Culture History. Because of its lack of interest in scientific evolution –descent with modification- the New Archeology did not restore these mistakes, and many of the current Darwinian Archaeologies inherited these defective conceptions (*e.g.* Ames 1996).

Conceiving the archaeological record as a fossil record, here I hold that a synthetic paradigm in evolutionary archaeology is necessary and convenient. A synthesizing framework is required not in order to suppress theoretical diversity, but as a common ground where a variability of competing hypotheses and models may proliferate, fueling the natural selection of scientific ideas. This requires a logical theoretical framework capable of linking the properties of the archaeological record, a distributional phenomenon of the present (see Dunnell 1992), with the complexities of the Darwinian theory of evolution.

In this work I advocate for the adoption of and extended theory of Darwinian evolution, by which natural selection is a general process which might occur on biological objects at many levels of a genealogical hierarchy, with several units changing along histories of decent with modification (Gould 1994, 2002). Artifacts are one of such evolving units (Neff 2001).

Upon these notions I argue that adaptationist and populational thinking, as in evolutionary biology, is the best avenue to modelling the selective environments where humans, artifacts and behaviours evolve by nested selective forces acting at multiple focal levels of a genealogical hierarchy. As natural selection produces populational level adaptations at particular focal levels, adaptationist and populational thinking are the common ground where evolutionary archaeology might converge in a synthetic paradigm. In this framework, by including some elements developed in the fields of evolutionary ecology and sociobiology, rewritten in proper archaeological terms, micro and macroevolutionary processes are not logically opposed. A unified Darwinism in evolutionary archaeology shall results from the more general extended theory of Darwinian evolution applied to the evolution of humans and other cultural animals. Hence, the unified Darwinian evolutionary archaeology paradigm will impact on the arguments of those scholars as Smith (2000) and Richerson and Boyd (2005) that by advocating for complementarities and for a synthetic theory of human behavior and evolution are reluctant to accept what evolutionary archaeology has demonstrated since its foundation: that natural selection directly acts on cultural variation as in genetic variation, without adopting a single-level reductionism and the artifices of a dual evolutionary model (*i.e.* Durham 1991).

EVOLUTIONARY ARCHAEOLOGY IS NOT A SINGLE-LEVEL REDUCTIONISTIC THEORY

Conceiving artifacts and behaviors as a part of the human phenotypes, evolutionary archaeology explains change in the archaeological record by the direct action of natural selection and other evolutionary processes on heritable variation (Dunnell 1980, 1989; Leonard and Jones 1987; Rindos, 1984, 1989; O'Brien and Holland 1990, 1992; O'Brien and Lyman 2000, 2002, 2003a, 2003b; Neff 2001; Neff and Larson 1997; Teltser 1995, O'Brien y Lyman 1998,).

Through a strong reliance on the phenotypic plasticity of modern humans, the whole research program of evolutionary archaeology has been questioned (see Boone and Smith 1998). Behind many of these critiques, it is a call to adopt the evolutionary ecology paradigm and the dual inheritance model of evolution (*e.g.* Ames 1996, Bettinger *et al* 1996, Bettinger and Richerson 1996, Boone and Smith 1998, Ortman 2001). These assertions are the logical derivations of following the reductionist uni-level selectionism characteristic of human behavioral ecology and sociobiology. The first advocates that individual organisms are the only valid units of selection (Smith and Winterhalder 1992), whereas for some versions of the later genes occupy this role (Dawkins 1976, 1982).

The issue about what actually evolves deserved the attention of the evolutionary paradigm since its foundation. By recognizing that artifacts and behaviours sometimes have differential inclusive fitness value, evolutionary archeologists often model evolutionary explanations taking the individuals as the unit of selection (O'Brien *et al* 1994). But as it is clear in the literature

evolutionary archaeology is not reductionist on this matter. For instance, Dunnell (1978, 1995) has suggested that cultural transmission creates an opportunity for the level of selection to shift up from the individual to groups, a hypothesis discussed on empirical grounds by Kosse (1994) and Shennan (2002, 2003).

Because of this, the term "replicative success" was introduced by Leonard and Jones (1987), asserting that the replicative success of a particular cultural trait might or might not affect the reproductive success of the bearer of such trait (Leonard and Jones 1987:216). Here become critical the notions of *replicator* – "a unit that passes on its structure directly through replication" -and *interactor* "an entity that directly interacts as a cohesive whole with its environment in such a way that replication is differentia (Hull 1980:318). Selection operating on populations of interactors in turn modifies the population level frequencies of replicators (Hull 1980). For example, replicators may consist of the information comprised in the instructions or *recipes of action to* make and use some material forms -artifacts (O'Brien and Lyman 2003). Therefore, by conceiving that artifacts are interactors, evolutionary archaeologists explicitly have assumed that selection often directly act at the level of the artifact, structuring populations of interacting artifacts (Lyman and O'Brien 1998, Neff 2001, O'Brien and Lyman 2000). For natural selection to occur, fitness differences must exist between the interactors of an evolving population. Hence, whereas reproductive success is a critical component of fitness, a general definition of fitness is not attached to reproduction. Rather, fitness is defined and measured in terms of successful information transmission (Barton 2008). In this way, by considering multiple interactors, the epistemology of evolutionary archaeology avoided a uni-level reductionist framework. A similar multilevel perspective on selection guided many other biological disciplines to define evolutionary units suited to the properties of their own empirical domains, such as evolutionary genetics (Merlo *et al* 2007), metapopulation biology and evolution (Hanski and Gilpin, 1997, Olivieri and Gouyon 1997) and paleobiology (Eldredge 1989). Hence, the issue at hand is how this hierarchical multilevel framework would look in evolutionary archaeology. In the next paragraph I shall briefly face this topic.

ARTIFACTS ARE EVOLVING UNITS IN A HIERARCHY OF EVOLVING INTERACTORS

Artifacts have five important properties by which they achieve evolutionary individuality (*sensu* Gould 2002): **First**: artifacts have descendents by external replication, leading to genealogical processes. **Second**: artifacts have discreet existence intervals, which mean that they began to exist and perish along time, having ontogeny. **Third**: artifacts vary with respect to the traits they posses. **Four**: the traits may impart different probabilities of survival and reproduction to the bearers, and differential replica-

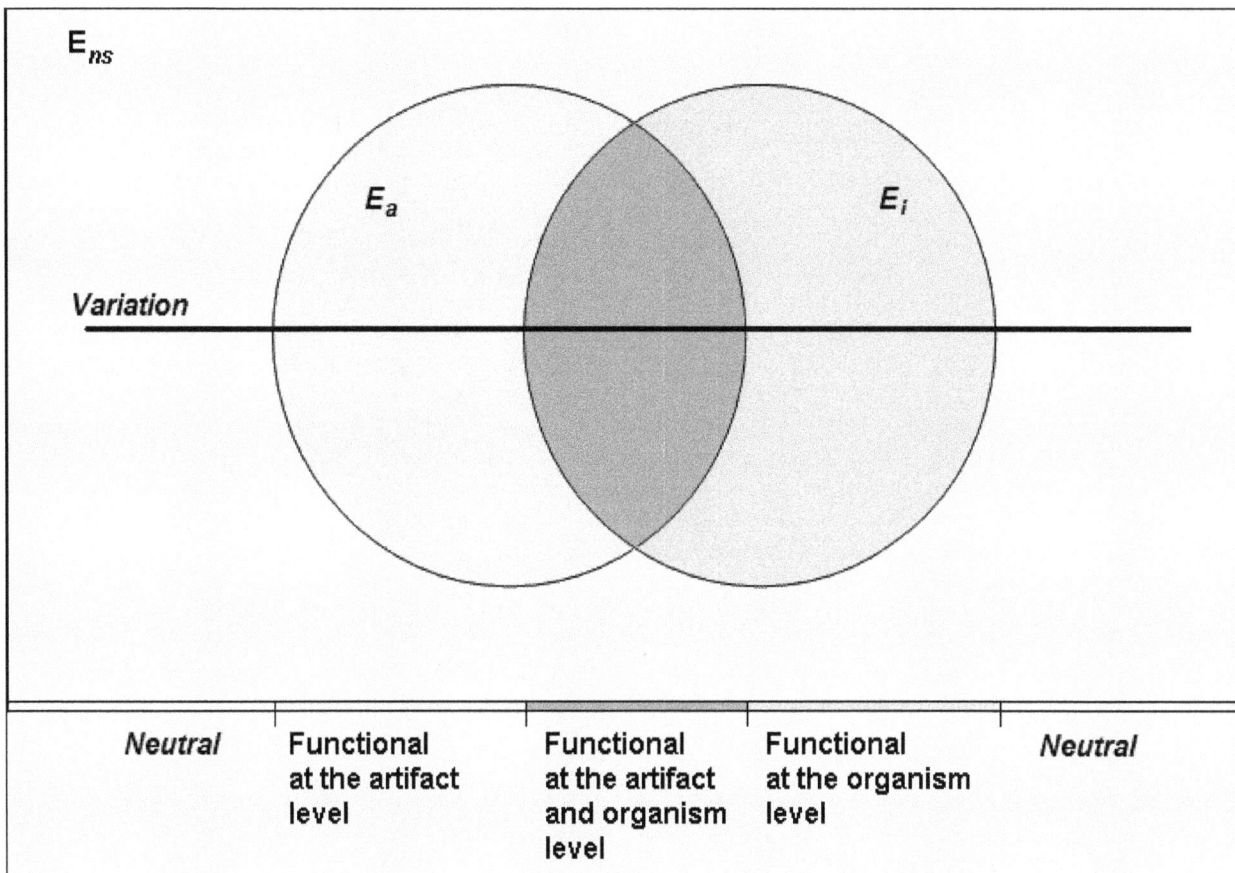

Fig. 10.1. Evolutionary environments and artifactual variation. Circle E_a represents the selective environment at the artifact scale; circle E_i represents the selective environment at the scale of the individual organisms, and E_{ns} is a non selective surface. Horizontal line represents artifactual variation. The labels show the status of the variation as a result of its interception with each environment (see text)

tive success to the artifacts into an artifactual linage or a cultural pool. *Five*; there is heritability between artifact forms, with great correlation between the traits of antecedent and descendant classes of artifacts resulted by transmission. Thereby, artifacts satisfy Lewontin´s (1970), and Gould (2002) criteria to define evolving individuals –interactors- by Darwinian selection.

As genuine selective units, artifacts are competing interactors in selective environments defined at their own scale. But when the interactors are more inclusive entities, as the individuals or groups bearing them, their selective evolution will result from the impact of the artifactual variation in the fitness of these entities. The logical conclusion of the above statements is that the functional status of the cultural variation comprised in artifacts is dependent on the focal level of the evolutionary process. This brings us the opportunity to build Darwinian adaptive explanations in a multilevel framework with different empirical expectations by modeling the different dimensions of the environments in which the variation is expected to be functional.

Figure 10.1 is a graphical representation of the expected status of the variation as a result of its expression along different evolutionary environments. Functional variation at the artifact level results when the variation is correlated with some dimension of the environment that only affects the replication and use of artifacts –the interception of the line with the E_a environment. Broadly, artifacts and technologies have *performances* to accomplish some tasks controlled by their designs (Schiffer and Skibo 1997). Accordingly, the evolutionary environments at the focal level of the artifact are shaped by forces acting on the base of the properties of the artifacts to be more or less attractive to potential users for performing certain tasks (Neff 2001). As artifact use and replication have associated costs in terms of time, energy, and knowledge, every one of these dimensions will impact over the differential probabilities of alternative artifactual variation of being successfully transmitted and replicated, producing selection.

Hence, when the selective evolution of cultural variation takes place only at the focal level of the artifact, the individuals manufacturing and using artifacts are agents of Darwinian selection. The selective forces at this scale emerge from the population level effects of what has been called "biased cultural transmission" (Boyd and Richerson 1985) plus other decision making mechanisms. As in other forms of selection, the empirical pattern of these processes is an *S* shaped representation of the variation along time, an slow-fast-slow frequency distribution (Bettinger 2008), with a temporal scale imposed by the replicative rate and life span of the artifacts and qualitative shorter than what is expected from selection acting upon organisms (Bettinger 2008).

When the artifactual variation is consistently correlated with the properties of the environment that affect the differential reproduction of the individual organisms it became functional at this evolutionary level -the interception of the solid line with E_i. The empirical signature of this process should be a slower evolutionary rate of the variation under selection imposed by reproductive rate and life span, with rates of evolution of archaeological resolution (see Laland and Brown 2006). Also, variation might be functional at more than one level, a situation represented in figure 10.1 by the line crossing the area of superposition of E_a and E_i. Shortly, artifacts and traits fixed by selection directly upon the artifact pool and having also positive biologically fitness effects became adaptations built by a non conflictive multilevel process. In addition to the short temporal scale sigmoidal distribution of the favored variant, the expected empirical expression of these processes is a demographical success correlated with the reproductive improvements of the individuals (see Edwards and O'Connell 1995). But when selection at infraorganismal level fixes variation that is deleterious for the individual organisms or groups, Darwinian selection at these upper levels will remove this variation (see Durham 1991).

The same logic is transferable to neutral variation (style). As figure 10.2 shows neutrality comprises archaeological variation not correlated with selective forces of *any* focal level. Because of this neutral variation has no detectable selective value (Dunnell 1978). The behavior along time of this variation results by drift and chance alone, according with the neutral model. Neutral evolutionary environments are governed by probabilistic events at several focal levels. For instance, at the level of the organisms, demographic stochasticity, migration and founder effect might fix neutral cultural variation in a population. Similar chance processes are expected at the level of the artifacts as results of the vagaries of transmission, sampling error, and recombination only taken place into a given cultural pool (Muscio 2004).

For selection to occur variation must be blind (*i.e.* innovation should be independent of selection). The extent to which variation in culture is blind is an empirical issue (Bettinger 2008, Mesoudi *et al* 2006). Nevertheless, at proper archaeological time scales what is expected is the dominance of blind variation and selection (Rindos 1984, 1989). Actually, our skills as selective agents of undirected variation occur all time in our every day life. For example, in shopping centers we are surrounded by an amazing amount of cultural variability, explicitly designed with the intention to bring some part of this

variability home. Artifact's designers, and brainstorm organizers, appeal to a multiplicity of devices to attract our attention, creating new artifact forms by the engineered recombination of existing variation, imitation, and innovation. But in a second step a crucial test happens: Darwinian selection. This is when variation confronts the nonlinear dynamics of the complex system emerging from collective individual behaviors (Barkley Rosser 2006). Confronting competition some artifacts have success and proliferate while others go directly to extinction (sometimes carrying also the firms to extinction). That intentional designers are incapable of predicting these nonlinear trajectories means that the variation they create is random with respect to natural selection, not only at the level of the artifacts, but at several focal levels in the hierarchy of interactors.

SEEKING COMMON ELEMENTS FOR A UNIFIED DARWINISM IN ARCHAEOLOGY

Behavioral ecology is a microevolutionary approach to phenotypic change along behavioral time (Smith 2000), whereas evolutionary archaeology is a macroevolutionary framework devoted to explain cultural change and the evolution of human populations along evolutionary times (Lyman and O'Brien 1998, 2001, Shennan 2003). Other microevolutionary approaches to cultural evolution are evolutionary memetics (Blackmore 1999), and the Cultural Transmission Theory (Cavalli Sforza and Feldman 1981, Boyd y Richerson 1985). Evolutionary archaeologists were always very aware that evolutionary ecology, sociobiology, and cultural transmission theory do not possess, *per se* any archaeological content.

Evolutionary archaeology and evolutionary ecology have little in common in terms of analytical goals, having different *explananda* (objects of explanation). But since the Darwinian research, in which both paradigms rely, demands populational and adaptive thinking (Dennett 1995), they have much in common in epistemological and methodological grounds. Not accidentally each paradigm decomposes complex phenomena in elementary pieces to explain. Behavioral ecology calls this strategy the *piecemeal approach* (Smith 2000), a way of explanatory reductionism (Winterhalder and Smith 1992). Evolutionary archaeology and paleobiology do the same under the label of *reverse engineering* (O'Brien *et al* 1994), where adaptive-selectionist thinking is the cornerstone to build arguments about function (Maxwell 2001).

Evolutionary archaeology has come to the conclusion that some of the ideas and methods built under the Culture History paradigm have critical explanatory value rewritten in Darwinian terms (Lyman *et al* 1997). Additionally, as Borrero (1993) suggested, explicitly materialistic Darwinian paradigms such as human behavioral ecology, might also have valuable elements, if rewritten in archaeological terms. Winterhalder and Goland (1996) Shennan (2003), Mesoudi (2008), among others, showed the utility of this approach to explain the emergence of large scale evolutionary changes through microescale evolutionary mechanisms.

More broadly, the unifying elements between human behavioral ecology, sociobiology, and evolutionary archaeology are the models predicting how people by adjusting their behavior to their current environments, in behavioral time, become selective agents of cultural variation, including artifacts and behaviors, remaining subject to natural selection at proper evolutionary times.

Predator-prey evolutionary dynamics exemplifies the issue at hand. At a populational level a predator –human or other animal- is a selective agent of its preys. Specialist predators obeying to their evolved design may modify the genetic composition of their preys, or by overexploitation or unstable local dynamics they may drive the local population of preys to extinction (Nee *et al.* 1997). In the last case predators may also go to extinction or to the evolution of new predator phenotypes by selection. Importantly, that predators may become extinct shows the limits of their evolved design into a new selective environment and over an evolutionary time scale. The same logic prevails when humans use resources and artifacts.

Humans, as agents of selection expressing their evolved nature, have the potential to modify the genetic and memetic information of their inherited environment, as *niche construction* predicts (Odling-Smee *et al.* 2003). Ecological inheritance is an important concept here, referring to the selective environment that new generations of individuals inherit from their ancestors, along evolutionary times (Odling-Smee *et al.* 2003). When humans, in behavioral times, introduce novel variation or act as agent of selection of preexisting variation they modify the selective environment for the next generations. The consequences of ecological inheritance can be archaeologically studied (Martínez 2002, Riede 2008).

As evolution may be conceived as an economic process in nature (Eldredge 1989), the models based on optimality have profound analytical value. Since energy and nutrients acquisition is critical for the survival of an organism during its whole ontogeny and reproduction, it is necessary to appeal to natural selection to explain the archaeological record of food consumption, including artifacts linked to food acquisition and consumption (Gremillion 2002).

Foraging behavior is a central issue for the application of adaptive models (Bettinger 1991). As humans socially learn many of their adaptive behaviors, including foraging strategies, what is expected is that cultural transmission, especially vertical transmission, takes the control of these behaviors (Guglielmino *et al.* 1995). Optimal diet models predict that humans will prefer higher return rate recourses; for example selecting big size animals instead of small game in particular environments, since the firsts

lead to higher return rates. Then, assuming cultural transmission, and taking the hunting practice as a phenotypic trait subject to lower level selection pressures, the individual adaptive bias of preferring big game will begin the selective retention, at the scale of the population, of the *big game hunting practice* - a cultural trait. Since evolution runs on non-behavioral times, natural selection will ship-up to the level of the individuals. Since the cultural variation, retained for selection at the scale of the behavior, is adaptive on the focal level of the individuals, a no conflicting nested multilevel selective process will retain the big game hunting practice against small game hunting, leading to the selective evolution of an *economic tradition*. This example illustrates the logic by which it is possible to link important behavioral models to complex evolutionary processes of large temporal scales. Therefore, if we are going to apply some selectionist ecological models, we must be aware that they will be useful not by assuming "optimal designs" expressed on behavioral-ethnographical times to use as interpretative devices; but for their capacity to predict in a probabilistic way (see Dunnel 1999) the kind of phenotypic variation that natural selection will retain over evolutionary time frameworks, and regarding specific socio-environmental variables. The critical point to remark is that because we are interested in evolution these models allow us to build hypotheses concerning some directional macroevolutionary patterns produced by humans acting as agents of selection over many populations of evolving interactors, but still subjected to natural selection.

AN EXAMPLE

Optimality models are useful tools to apply in reverse engineering analyses, generating hypotheses about the context-dependent functional dimension of the variation under some optimal criteria. For instance, the puna region of Northwestern Argentina is a high altitude desert located at 3.000 meters above the sea level. Resource availability to use as fuel is scarce. In addition, hypoxia (a low concentration of atmospheric oxygen because of a higher altitude) increases the requirements of fuel for cooking. Ethnographical and experimental research conducted among farmers of the puna shows that cooked resources return rates are highly conditioned by fuel availability and cooking technologies, particularly corn return rates (Muscio 2004). Likewise, ceramic production costs are highly impacted by fuel availability (Camino 2006). From this knowledge about the selective environment of small scale pottery production and use in the puna, it was proposed that natural selection would fix, in a ceramic linage, any trait with the potential to diminish the costs of cooking and the costs of vessels replacement, enhancing the return rate of cooked resources (Muscio 2004). In the puna of Argentina the first ceramic record is dated around 3.000 Bp, associated with herding and hunting economies. Later, around 2.500 Bp, ceramics appear associated with agricultural and

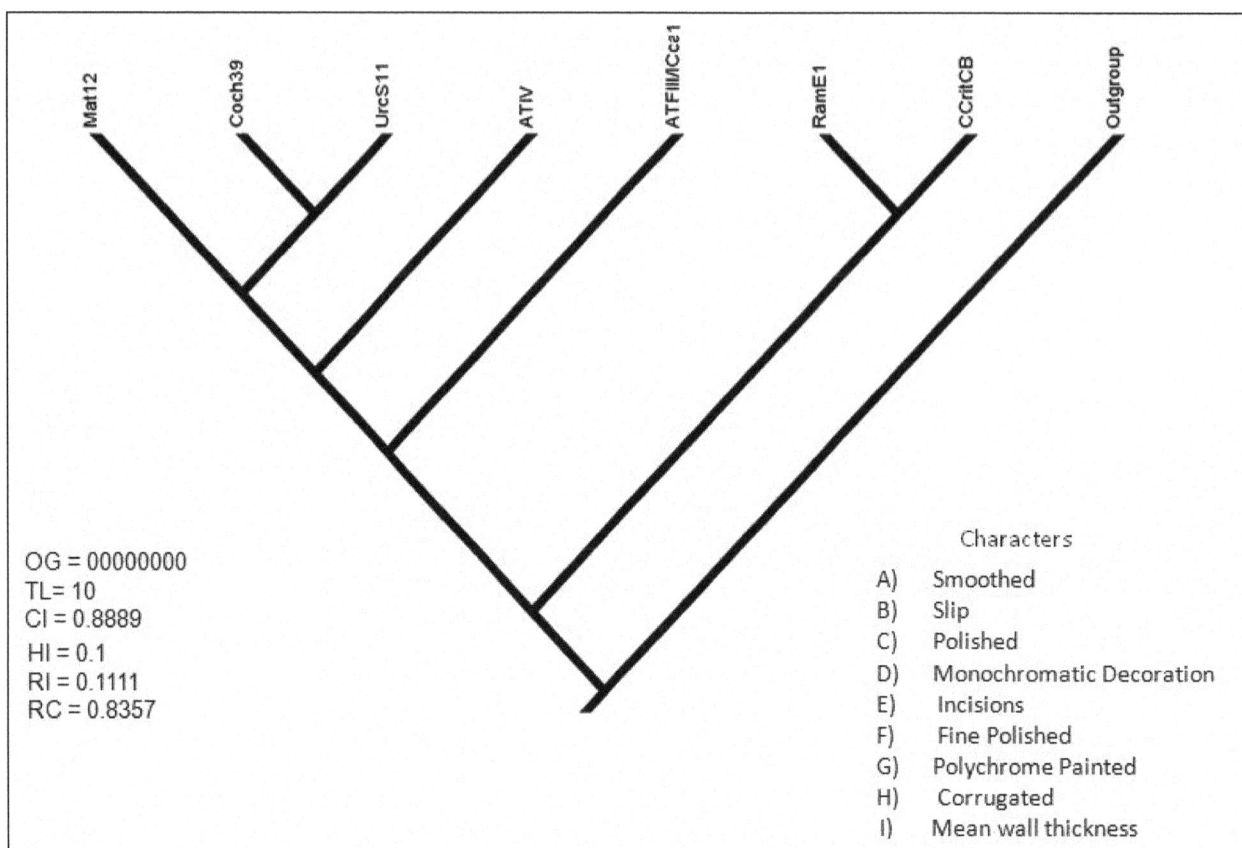

Fig. 10.2. Cladogram of the earliest ceramics of Northwestern Argentina. Each terminal taxa is a single assemblage. Mat12: Matancillas 1 and 2; Coch39: Cochinoca 39; UrcS11: Urcuro Sondeo 11; ATFIII: Alero Tomayoc Fase III; ATIV: Alero Tomayoc Fase IV; Ica1: Inca cueva Alero1; CcritCB: Cueva Cristobal Capa B; RE1: Ramadas Estructura. The information was obtained from the published bibliography (see Muscio 2004). Character states are binary, except for mean wall thickness

herding economies. The genealogical side of the hypothesis was investigated with cladistics and occurrence seriation analyses (O'Brien and Lyman 2001). Selection was assessed investigating the behavior along time of a single continues character: mean wall thickness, which in cooking vessels is positively correlated with thermal conductivity and thermal stress resistance (O'Brien *et al* 1994). Figure 10.2 shows the rooted cladogram of the earliest ceramic aggregates of the region. Only one cladogram was retained from an exhaustive search using PAUP 4.1 (Swofford 2002) from a matrix of eight binary characters, including mean wall thickness (see details in Muscio 2004). The phylogenetic signal is strong and significant (CI=0.8889, RI=0.8357, TL=10), documenting branching evolution. Figure 10.3 shows a robust pattern of declination along time of mean wall thickness, with a fast rate of –0.001 mm/year, which is consistent with a directional optimizing selection processes. This rate of change can be interpreted as the strong selective control of decision making and biased transmission forces over the wall thickness of the vessels. In a context of spatial aggregation and mobility reduction associated with food production, thin wall vessels, by enhancing the return rates of cooked resources should have enhanced the reproductive success of pottery users. Hence, independent

evidence is needed to test the demographical implications of this hypothesis.

CONCLUSION

Dennet (1996) suggested that evolutionary ecology and evolutionary archaeology are entirely compatible, arguing that adaptive-directed selection of cultural variation is a *variety* of natural selection. In agreement with this idea, here I proposed that this variety of selection is a consequence of the change in the level of the evolving interactors, from the individual organism to the artifacts and behaviors. In short, the links between microevolutionary mechanisms to macroevolutionary patterns come from the realization that cultural variants usually evolve in environments that include a particularly well-focused selective pressure consisting of human agents, and that these agents are not free from the action of selection .

Upon this basis a unified paradigm in evolutionary archaeology is possible and convenient. It is possible because natural selection as a mechanism is not confined to only one kind of interactor, and the archaeological

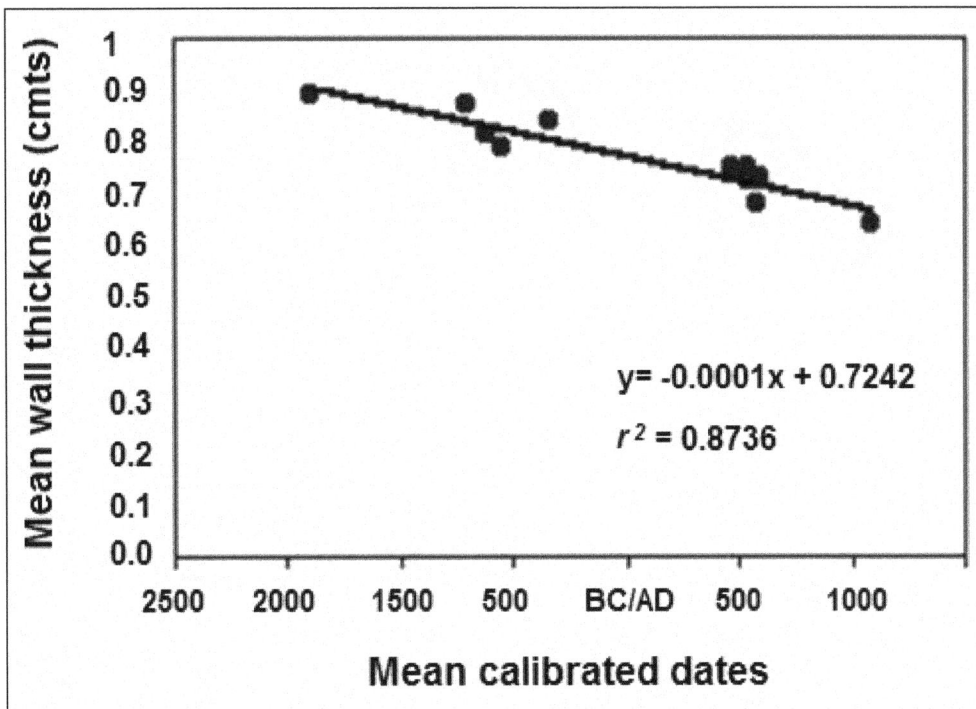

Fig. 10.3. Declination throughout time of the mean wall thickness of the earliest ceramics of Argentina

record documents the evolutionary history of many interactors. It is convenient because evolutionary archaeology needs to enhance the inclusiveness of phenomena to explain with better learning strategies to account for the past than those derived from a single level reductionist approach.

Science demands the existence of standards to assess the true content of competing hypotheses, in a process of selective retention of ideas. As in evolutionary biology, a synthetic paradigm in evolutionary archaeology will provide the theoretical basis for such standards. In this way, an expanded synthetic Darwinism in evolutionary archaeology increases the inclusiveness of the paradigm, leading to the integration of several lines of selectionist research on human behavior and evolution into a single logical framework rooted in evolutionary biology.

Acknowledgements

I thank to Marcelo Cardillo and Gabriel López for providing valuable comments on an early draft of this essay. This work was supported by CONICET, Consejo Nacional de Investigaciones Científicas y Técnicas de Argentina.

References

AMES, K.M. 1996 Archaeology, style, and the theory of coevolution. In Darwinian Archaeologies, p. 109-131. H.G. Maschner (ed.), Plenun Press, New York.

BAMFORTH, D.B. 2002. Evidence and metaphor in evolutionary archaeology. American Antiquity (67: 435–52).

BARTON, C. M. 2008. General fitness, transmission, and human behavioral systems. In Cultural Transmission and Archaeology Issues and Case Studies, p. 112-119, Michael J., O'Brien (ed.). Society for American Archaeology.

BARKLEY ROSSER, J. 2006. From catastrophe to Chaos: A General theory of Economic Discontinuities: Mathematics, Microeconomics, Macroeconomics, and Finance (Volume I) (Mathematics, Microeconomics and Finance). Kluwer Academics Publishers.

BETTINGER, R.L. 1991. Hunter-gatherers: Archaeological and Evolutionary Theory. Plenum Press. New York.

BETTINGER, R.L. (2008) Cultural transmission and archaeology. In Cultural Transmission and Archaeology Issues and Case Studies, p. 1-9, Michael J, O'Brien (ed.). Society for American Archaeology.

BETTINGER, R.; BOYD, R.; RICHERSON, P.J. 1996. Style, function and cultural evolutionary processes. In Darwinian Archaeologies, p 133-164. H.G. Maschner (ed.), Plenun Press, New York.

BETTINGER, R.; RICHERSON, P. J. 1996. The state of evolutionary archaeology: Evolutionary correctness, or the search for the common ground. In Darwinian Archaeologies, p 221-231. H.G. Maschner (ed.). Plenun Press, New York.

BLACKMORE, S. 1999. The Meme Machine. Oxford University Press.

BOONE, J. L.; SMITH, E.A. 1998. Is it evolution yet? A critique of "Evolutionary Archaeology", Current Anthropology (39:141-173).

BORRERO, L. 1993. Artefactos y evolución. Palimpsesto Revista de Arqueología (3:15-32). Buenos Aires.

BOYD, R.; RICHERSON, P.J. 1985. Culture and the Evolutionary Process. Chicago: University of Chicago Press.

CAMINO, U. 2006. La cerámica del Período Agro-Alfarero temprano de la quebrada de Matancillas (puna de la Provincia de Salta). Tesis de licenciatura en Antropología. FFyl, Universidad de Buenos Aires. Ms.

CAVALLI-SFORZA, L.L.; FELDMAN, M.W. 1981. Cultural Transmission and Evolution. Princeton: Princeton University Press.

CULLEN, B. 1996. Cultural Virus Theory: Archaeology and the Nature of Construction. Cambridge University Press.

DAWKINS, R. 1976. The Selfish Gene. Oxford University Press. Oxford.

DAWKINS, R. 1982. The Extended Phenotype: The Gene As The Unit of Selection. Oxford University press, Oxford.

DENNETT, D.C. 1995. Darwin's Dangerous Idea. Evolution and the Meanings of Life. Penguin Books. London.

DENNETT, D.C. 1996. Comments on: Is it evolution yet? A critique of "Evolutionary Archaeology", by J.L. Bonne and E.A. Smith, Current Anthropology (39: 141-173).

DUNNELL, R.C. 1978. Style and function: A fundamental dichotomy, American Antiquity (43:192-202).

DUNNELL, R.C. 1980. Evolutionary theory and archaeology. Advances in Archaeological Method and Theory, 3, 35-99.

DUNNELL, R.C. 1995. What is it that actually evolves? In Evolutionary archaeology: Methodological issues, p 33-50. Patrice A. Teltser (ed.), Tucson: University of Arizona Press.

DUNNELL, R.C. 1999. The Concept of Waste in an Evolutionary Archaeology, Journal of Anthropological Archaeology (18, 243–250).

DUNNELL, R.C. 1992. The notion site. Space, time, and archaeological landscapes, p 21-41. Rossignol, J.; Wandsnider, L. (eds.) Plenum Press, New York.

DUNNELL, R.C. 2001. Foreword. In: Style and Function: Conceptual Issues in Evolutionary Archaeology, (xiiv-xxiv). Hurt T. and G. Rakita, (eds.). Bergin and Garvey, Westport. Connecticut.

DURHAM, W. 1991. Coevolution: Genes, Culture and Human Diversity. Stanford University Press, Stanford.

EDWARDS, D.; O'CONNELL, J.F. 1995. Broad Spectrum Diets in Arid Australia. Antiquity 69:265.

ELDREDGE, N. 1989. Macroevolutionary Dynamics: Species, Niches and Adaptive Peaks. MacGraw-Hill, New York.

FOLEY, R. 1992. Evolutionary ecology of fossil hominids. In: Evolutionary Ecology and Human Behavior, p. 131-164. Smith E.A; Winterhalder, B. (eds.). Aldine de Gruyter, New York.

GOULD, S.J. 1994. Tempo and mode in the macroevolutionary reconstruction of Darwinism, Proceedings of The national Academic of Sciences of the Unites States of America (Vol. 91:6764-6771)

GOULD, S.J. 2002. The Structure of Evolutionary Theory. The Belknap Press of Harvard University Press.

GREMILLION, K.J. 2002. Foraging theory and hypothesis testing in archaeology: An exploration of methodological problems and solutions. Journal of Anthropological Archaeology (21:142–164).

GUGLIELMINO, C.R; VIGANOTTI, C: HEWLETT, B; CAVALLI-SFORZA, L.L. 1995. Cultural Variation in Africa: Role of mechanisms of transmission and adaptation. Proceedings of The National Academic of Sciences of the Unites States of America (92:7585-7589).

HANSKI I., A; GILPIN, M.E. 1997. Metapopulation Biology: Ecology, Genetics, and Evolution. Elsevier Science, Academic Press.

HULL, D. 1980. Individuality and selection. Annual Review of ecology and Systematics (1:1-18).

HURT, T.; VAN POOL, T.L.; LEONARD, R.D.; RAKITA, G. 2001. Explaining the co-occurrence of attributes in the archaeological record: A further consideration of replicative success. In: Style and Function: Conceptual Issues in Evolutionary Archaeology, p. 51-68. Hurt T. and G. Rakita, (eds.). Bergin and Garvey, Westport. Connecticut.

LALAND, K.N.; BROWN, G.R. 2006. Niche construction, human behavior, and the Adaptive-Lag Hypothesis. Evolutionary Anthropology, 15:95-104.

LEONARD, R.D; JONES G.T. 1987. Elements of an Inclusive Evolutionary Model for Archaeology. Journal of Anthropological Archaeology 6: 199-219.

LEWONTIN, R.C. 1970. The units of selection. Annual Review of Ecology and Systematics (1:1-18).

LYMAN, R.L. 2001. Culture historical and biological approaches to identifying homologous traits, In: Style and Function: Conceptual Issues in Evolutionary Archaeology, p. 68-89. Hurt T. and G. Rakita (eds.). Bergin and Garvey, Westport. Connecticut.

LYMAN, R.L.; O'BRIEN, M.J. 1998. The goals of evolutionary archaeology: History and explanation. Current Anthropology, 39, 615-652.

LYMAN, R.L.; O'BRIEN, M.J. 2001. On misconceptions of evolutionary archaeology: Confusing macroevolution and microevolution, Current Anthropology, (42: 408-409).

LYMAN, R.L.; O'BRIEN, M.J. 2003. Cultural traits: Units of analysis in early twenty-century anthropology. Journal of Anthropological Research, 59: 225-250.

LYMAN, R.L.; O'BRIEN M.J; DUNNELL R.C. 1997. Americanist Culture History Fundamentals of Time, Space, and Form. Plenum Press, New York.

MARTÍNEZ, G. 2002. Organización y cambio en las estrategias tecnológicas. Un caso arqueológico e implicaciones conductuales para la evolución de las sociedades cazadoras-recolectoras pampeanas. In: Perspectivas Integradoras en Arqueología y Evolución: Teoría, Métodos y Casos de Aplicación, G.A. Martínez y J.L. Lanata (eds.): 286-301. INCUAPA, Olavarría.

MASCHNER, H.D.G. 1996. Darwinian Archaeologies. Plenum Press, New York.

MAXWELL T.W 2001. *Directionality, function, and adaptation in the archaeological record.* In: Style and Function: Conceptual Issues in Evolutionary Archaeology, p. 41-50. Hurt T. and G. Rakita (eds.). Bergin and Garvey, Westport. Connecticut.

MESOUDI, A, 2008. The experimental study of cultural transmission and its potential for explaining archaeological data. In Cultural Transmission and Archaeology Issues and Case Studies, p. 91-101. Michael J, O'Brien (ed.), Society for American Archaeology.

MERLO, L.M., F. PEPPER, J.W. REID, B.J; MALEY, C.C. 2006. Cancer as an Evolutionary and Ecological Process. Nature Reviews Cancer 6, 924-935.

MESOUDI, A.; WHITEN, A.; LALAND, K.N. 2006. Towards a unified science of cultural evolution. Behavioral and Brain Sciences. 29:(4) 329-383.

MUSCIO, H.J. 2002. Cultura material y evolución. In: Perspectivas Integradoras en Arqueología y Evolución: Teoría, Métodos y Casos de Aplicación, p. 21-54. Martínez, G.A.; Lanata J.L. (eds.). INCUAPA, Olavarría.

MUSCIO, H.J. 2004. Dinámica Poblacional y Evolución Durante el Período Agroalfarero Temprano en el Valle de San Antonio de los Cobres, Puna de Salta, Argentina. Doctoral thesis, FFyL, Universidad de Buenos Aires, Argentina.

NEE, S.; MAY, N.; HASSEL, M.P. 1997. Two-species metapopulation models. In: Metapopulation Biology: Ecology, Genetics, and Evolution, p. 123-148. Hanski I., A; Gilpin, M.E (eds.). Elsevier Science, Academic Press.

NEFF, H. 2000. On evolutionary ecology and evolutionary archaeology: Some common ground, Current Anthropology (41:427–429).

NEFF, H. 2001. Differential persistence of what? The scale of selection issue in evolutionary archaeology. In: Style and Function: Conceptual Issues in Evolutionary Archaeology, p. 25-40. Hurt T. and G. Rakita (eds.). Bergin and Garvey, Westport. Connecticut.

NEFF, H., LARSON, D. 1997. Methodology of comparison in evolutionary archaeology. In: Rediscovering Darwin: Evolutionary Theory and Archeological Explanation, p. 75-94. Barton, C.M. Clark, G.A (eds.). American Anthropological Association, Washington, D.C.

O'BRIEN, M.J.; LYMAN, R.L. 2000. Applying evolutionary archaeology. New York: Kluwer Academic.

O'BRIEN, M.J.; LYMAN, R.L. 2002. Evolutionary archeology: Current status and future prospects. Evolutionary Anthropology, 11, 26-36.

O'BRIEN, M.J.; LYMAN, R.L. 2003a. Cladistics and archaeology. Salt Lake City: University of Utah Press.

O'BRIEN, M.J.; LYMAN, R.L. 2003b. Resolving phylogeny: Evolutionary archaeology's fundamental issue. In: Essential Tensions in Archaeological Method and Theory, p.115-135. VanPool T.L, VanPool. C.S. (eds.) Salt Lake City: University of Utah Press.

O'BRIEN M.J.; HOLLAND, T.D. 1990. Variation, selection, and the archaeological record. Studies in Archaeological Method and Theory, Vol. 2, (31-79) Schiffer, M.B (ed.). Tucson: University of Arizona Press.

O'BRIEN M.J.; HOLLAND, T.D. 1992. The Role of Adaptation in Archaeological Explanation. American Antiquity (57: 36-59).

O'BRIEN, M. J.; HOLLAND, T.D.; HOARD, R.J; FOX, G.L. 1994. Evolutionary implications of design and performance characteristics of prehistoric pottery. Journal of Archaeological Method and Theory 1(3):211-304.

ODLING-SMEE F.J: LALAND K-N; FELDMAN M.W. 2003. Niche Construction: The Neglected Process in Evolution. Monographs in Population Biology 37. Princeton: Princeton University Press.

OLIVIERI, I.; GOUYON P.H. 1997. Evolution of migration rates and other traits: the metapopulation effect. In: Metapopulation Biology: Ecology, Genetics, and Evolution, p. 233-324. Hanski I., A; Gilpin, M.E. (eds.). Elsevier Science, Academic Press.

ORTMAN, S.G. 2001. On a fundamental false dichotomy in evolutionary archaeology: Response to Hurt, Rakita, and Leonard, American Antiquity (66-4:744-746).

KOSSE, K. 1994. The Evolution of large, complex groups: A hypothesis. Journal of Anthropological Archaeology (13:35-50).

RICHERSON, P.J.; BOYD, R. 2005. Not by Genes Alone: How Culture Transformed Human Evolution. Chicago: University of Chicago Press.

RIEDE, F. 2008. Maglemosian memes: Technological ontogeny, craft traditions, and the evolution of Northern European Barbed Points. In Cultural Transmission and Archaeology Issues and Case Studies, p. 178-189, Michael J., O'Brien (ed.), Society for American Archaeology.

RINDOS, D. 1984. The Origins of Agriculture: and Evolutionary Perspective. New York: Academic Press.

RINDOS, D. 1989. Darwinism and Its Role in the Explanation of Domestication, In: Foraging and Farming, The Evolution of Plant Domestication, p. 26-54. Harris D.R.; Hillman G (eds.) Unwin Hyman, Londres.

SHENNAN, S.J. 2002. The Darwinian archaeology of social norms and institutions: Issues and examples. In: Perspectivas Integradoras en Arqueología y Evolución: Teoría, Métodos y Casos de Aplicación, p. 157-174. Martínez, G.A.; Lanata J.L. (eds.). INCUAPA, Olavarría.

SHENNAN, S.J. 2003. Genes, Memes and Human History. London: Thames and Hudson.

SCHIFFER, M.B.; SKIBO, J.M. 1997. The explanation of artifact variability. American Antiquity 62 (1), 27-50.

SMITH, E.A. 2000. Three styles in the evolutionary analysis of human behavior. In: Adaptation and Human Behavior An Anthropological Perspective, p.27-46. Cronk, L; Chagnon N, Irons, W. (eds.). Aldine de Gruyter, New York.

SMITH, E.A.; WINTERHALDER, B. 1992. Evolutionary Ecology and Human Behavior. Aldine de Gruyter. New York.

STEELE, J. 2002. Evolution, ecology and human adaptability. In: Perspectivas Integradoras en Arqueología y Evolución: Teoría, Métodos y Casos de Aplicación, p. 207-219. Martínez, G.A.; Lanata J.L. (eds.). INCUAPA, Olavarría.

SWOFFORD, D.L., 2002. PAUP *: Phylogenetic Analysis Using Parsimony (and Other Methods). Sinauer Associates, Sunderland, 2002.

TELTSER, P.A. 1995. Evolutionary Archaeology, Methodological Issues, The University of Arizona Press, Tucson.

WILSON, D.S.; SOBER, E. 1994. Reintroducing group selection to the human behavioral sciences. Behavioral and Brain Sciences 17 (4): 585-654.

WINTERHALDER, B.; GOLAND. C. 1997. An evolutionary ecology perspective on diet choice, risk, and plant domestication. In: People, Plants, and Landscapes. Studies in Paleoethnobotany, (123-160). Gremillion, K.J., (ed.). The University of Alabama Press.

NICHE CONSTRUCTION APPLIED: TRIPLE-INHERITANCE INSIGHTS INTO THE PIONEER LATE GLACIAL COLONIZATION OF SOUTHERN SCANDINAVIA

Felix RIEDE

Leverhulme Centre for Human Evolutionary Studies, University of Cambridge,
The Henry Wellcome Building, Fitzwilliam Street, Cambridge, CB2 1QH, UK

Abstract: *In recent years, the niche construction model Odling-Smee et al. (2003) has gained increasing ground as a coherent and productive new research framework for understanding the co-evolution of genes, environments and behaviours. Building on the insights of Lewontin (2000) regarding the interactive nature of adaptation and the dual-inheritance approach for linking genetic with behavioural evolution (Richerson & Boyd 2005), this approach adds a third domain of inheritance: ecological inheritance. Using the pioneer Late Glacial re-colonization of Southern Scandinavia as a case study, this paper makes the case that the archaeological record provides much evidence for past human ecological inheritance.*
Key words: *Niche construction, evolutionary archaeology, Late Glacial, Scandinavia*

Résumé: *Récemment, le modèle d'Odling-Smee et al. (2003) de la construction de niche a gagné du terrain comme nouvelle approche d'une recherche cohérente et productive pour comprendre la co-évolution des gènes, des milieux et des comportements. Construite sur les idées de Lewontin (2000) à propos de la nature interactive de l'approximation de l'adaptation et de l'héritage duel pour mettre en relation génétique et évolution comportementale (Richerson & Boyd 2005), cette démarche ajoute un troisième domaine à l'héritage : l'héritage écologique. Utilisant comme cas d'étude la re-colonisation au Tardiglaciaire de la Scandinavie du Sud-est, ce travail en vient à la conclusion que le registre archéologique fournit des preuves à l'appui de l'héritage écologique du passé humain.*
Mots clés: *Construction de niche, archéologie évolutionniste, Tardiglaciaire, Scandinavie*

INTRODUCTION

Minimally, niche construction can be defined as the process through which organism modify their own, and often other organisms, environment to such a degree that they change the selective pressures acting on themselves as well as, importantly, their offspring (Fig. 11.1). While animals such as the beaver (genus *Castor*) are well known for their ability to substantially modify their immediate surroundings, Odling-Smee *et al.* (2003) provide a long list of other organisms – ranging from the lowly earthworm to humans – whose environmental modifications have impacted or do impact directly on the course of their genetic evolution. Putting this 'neglected process' in evolution in its rightful place amongst selection and drift is increasingly "causing a commotion" (Laland *et al.* 2004: 609) in theoretical evolutionary biology, but it has also stimulated a productive research program in which issues of adaptation (Day *et al.* 2003), ecology (Laland *et al.* 1999; Odling-Smee *et al.* 1996) and cultural evolution (Laland & Brown 2006; Laland *et al.* 2000, 2001; Odling-Smee 2006) are being reconsidered.

Despite the tremendous effort that the champions of niche construction have put into documenting and defending their approach (Laland & Sterelny 2006; Odling-Smee *et al.* 1996, 2003), they have all but neglected the archaeological record. However, Sterelny (2006: 151-152) has recently noted that "to the extent that information does flow collectively, niche construction is our best model of the generation-by-generation accumulation of skill, technology and information". He rightly stresses the importance of distinguishing between collectively held

and used information or technology and personally held and used technology and information. The latter falls squarely into the domain of cultural inheritance, whereas the former reflects ecological inheritance. To-date, most evolutionary archaeological case studies that address long-term material culture change focus on technologies that are effectively based on personal rather than collective knowledge, such as projectile points or ceramics (Collard & Shennan 2000; Collard & Tehrani 2005; Croes *et al.* 2005; Jordan & Mace 2006; O'Brien *et al.* 2001; O'Brien *et al.* 2002). Critically, however, the distinction between collective and personal knowledge echoes Oswalt's (1976) categorization of technologies into (tended and untended) *facilities* and *instruments/weapons* respectively. This paper then argues that the relationship between archaeology and the niche construction model is reciprocal: Archaeology provides information on exactly how humans became such powerful niche constructors, and the niche construction model in turn provides and overarching Darwinian rationale for including a much wider range of archaeological data than has hitherto been possible. Here, the Late Glacial pioneer re-colonization of Southern Scandinavia (Fig. 11.2) is examined from a niche construction perspective. While the legacy of ecological modifications of foragers is likely to be subtle, colonization scenarios are ideally suited for investigating the human impact on uninhabited landscapes. Current hypotheses regarding this re-colonization process range from migrationist (e.g., Rust 1942) to adaptationist (e.g., Fischer 1989a), but none have so far been tested. Yet, "northern Europe is an extraordinary laboratory for the investigation of human colonization and adaptation" (Price 1991: 185) and there

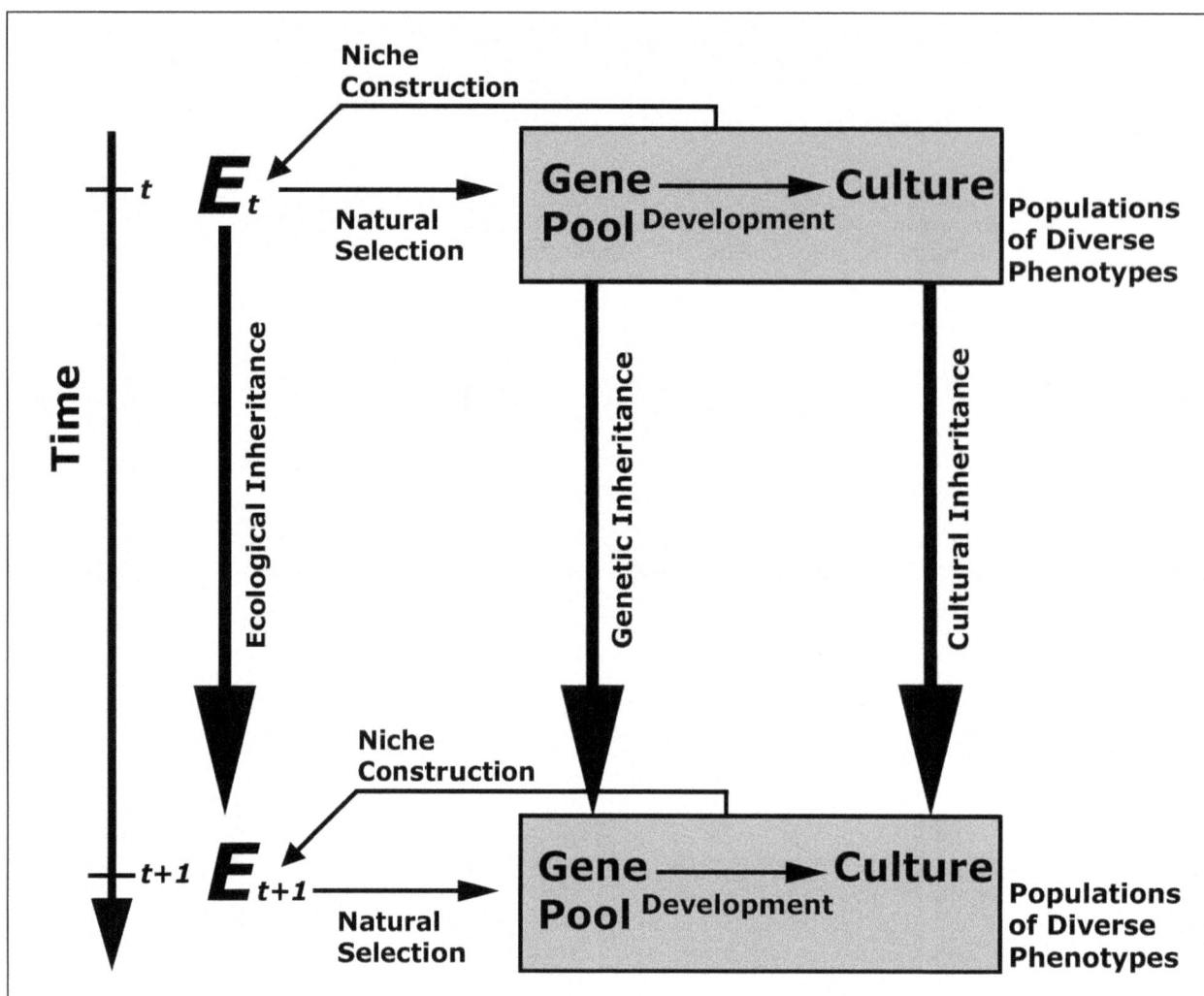

Fig. 11.1. A schematic outline of the niche construction or triple-inheritance model. The total phenotype of the members of the population at t is composed of genetic and cultural components. These are connected to the succeeding descendant generation at t+1 through genetic and cultural information transmission. This part of the model is identical to dual-inheritance models of bio-social evolution. The niche construction approach, however, also recognizes a third domain on inheritance, ecological inheritance. At t, organisms from the population modify their environment so that the selection pressures exerted on them at reproduction are modified. This modified environment in then bequeathed onto the descendant generation at t+1 through ecological inheritance

is no approach better suited to testing such issues than the Darwinian one. The three domains of inheritance are considered in turn.

GENETIC INHERITANCE

Because skeletal data from pre-Holocene periods in Southern Scandinavia is missing entirely (Newell *et al.* 1979), investigating modern genetic data – the selectively neutral maternally inherited mtDNA and the paternally inherited Y-chromosome – and using phylogenetic methods of reconstructing past population dynamics remains the only option (Cavalli-Sforza & Feldman 2003; Goldstein & Chikhi 2002; Jobling *et al.* 2004). Previous work in this field has used relatively crude summary statistic measures to distinguish populations, but it is the

phylogeographic approach that has proven particularly useful (Sykes 2000). This approach focuses on the spatial distribution of particular clades or haplogroups that have been dated via the molecular clock to the period of interest. Using phylogeographic methods, workers have been able to demonstrate that around 70% of the modern European mtDNA gene pool derives from various Palaeolithic expansion episodes and not as previously thought from the Neolithic settlement by agriculturalists from the Near East (Forster 2004).

Whole-genome sequencing of human mitochondria has provided further support to this hypothesis. It has been possible to distinguish at least two if not more colonization episodes with in the most common haplogroup (H) alone (Achili *et al.* 2004; Loogväli *et al.* 2004; Pereira *et al.* 2005). Molecular clock dating puts these episodes into

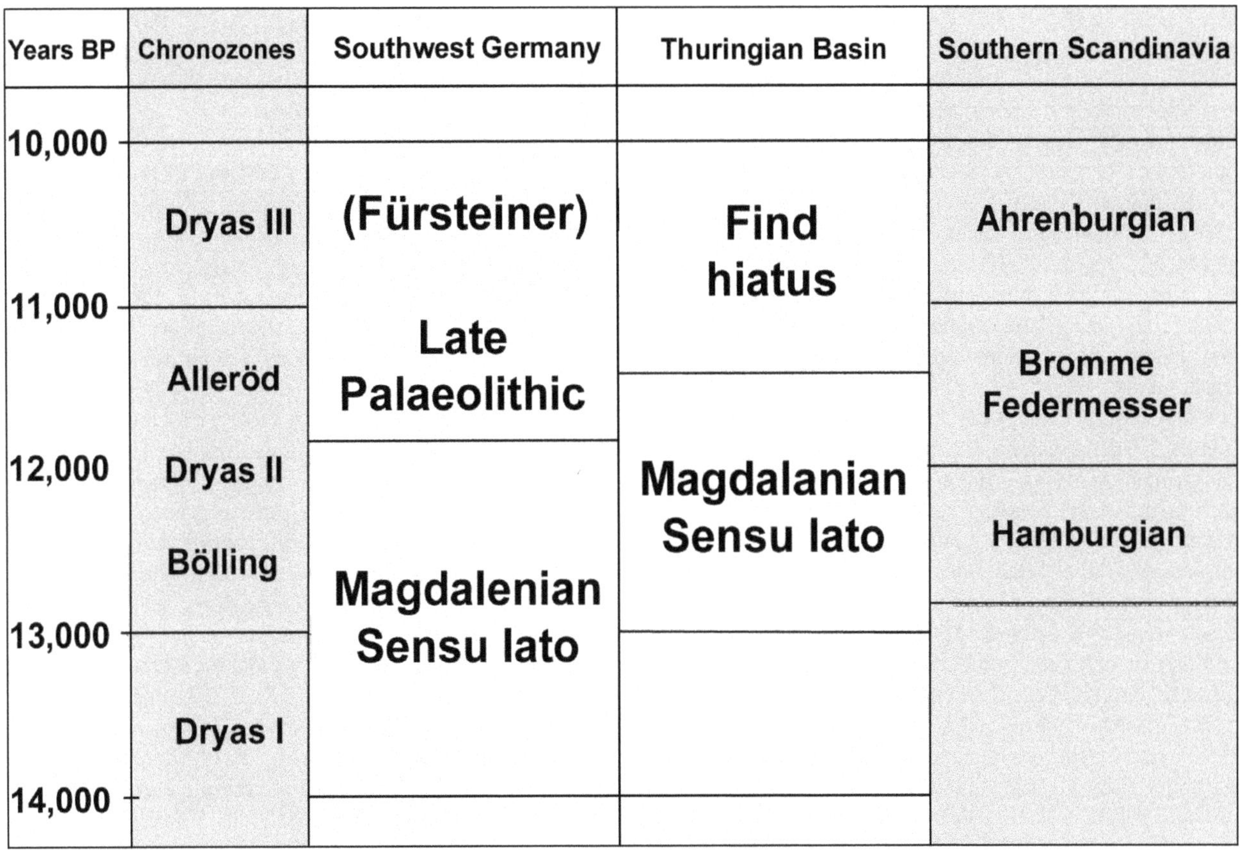

Years BP	Chronozones	Southwest Germany	Thuringian Basin	Southern Scandinavia
10,000				
	Dryas III	(Fürsteiner)	Find hiatus	Ahrenburgian
11,000		Late Palaeolithic		
	Alleröd			Bromme Federmesser
12,000	Dryas II		Magdalanian Sensu lato	
	Bölling	Magdalenian Sensu lato		Hamburgian
13,000				
	Dryas I			
14,000				

Fig. 11.2. A schematic outline of the Late Glacial in northern and central Europe. Adapted from Eriksen (1996). The three regions shown here correspond to those in Figure 11.6

the far side of the Holocene-Pleistocene border, a period during which short-lived climatic ameliorations were followed by phases of severe arctic conditions. Low population densities coupled with extremely rapid and pronounced climate change would have made peripheral groups rather prone to local extinction.

The Y-chromosome complements this picture with the male perspective. Severe bottlenecks producing the current Y-chromosome configurations in Scandinavia (Karlsson et al. 2006; Passarino et al. 2002; Sajantila & Pääbo 1995; Sajantila et al. 1996) have been identified. Passarino et al. (2002: 424) specifically argue that "at least from the male perspective, the genetic pool of the Norwegians is mainly composed of genes that were present in Europe as early as the Palaeolithic" and Tambets et al. (2004: 678) add that "this genetics-based reconstruction...is in agreement with the reconstruction of the spread of Ahrensburgian and Swiderian Mesolithic technologies in northern Europe, linking it with population expansion that can be likely traced back to the post–Last Glacial Maximum recolonization of the European north". This suggestion begs the question of what happened between the very first colonization some 14,500 BP and the eventual emergence and spread of the Ahrensburgian and Swiderian cultures into northern Scandinavia c. 2.500 years later (Fuglestvedt 2005; Kindgren 2002; Schmitt et al. 2006).

Genetic inheritance provides insights into the biological inheritance dynamics during the Late Glacial. Although great advances have been made in recent years regarding the resolution of this data, the large standard errors associated with molecular clock dating make it difficult to link particular expansion episodes to a given archaeological signature. Nevertheless, the emerging picture form genetics highlights the possibility of demographic instability during the Late Glacial, and it is with this in mind that one can turn towards ecological inheritance.

ECOLOGICAL INHERITANCE

Much archaeological data directly relates to ecological inheritance. It is trivial to say that culture is the human niche (Hardesty 1972), but it highlights the fact that human practice such extensive ecosystem engineering that many facets of human culture relate only to aspects of their niche that they themselves created. As Richerson & Boyd (2005) point out there are many adaptive peaks in the landscape of cultural evolution and that actors have very little means of evaluating the long-term evolutionary consequences of any given behaviour. Risk and adaptedness are to some degree culturally constituted (Fitzhugh 2001; van der Leeuw 1989) and cultural inertia can maintain certain practices despite being mal-adaptive. Making the correct adaptive choices would have been made more difficult still by rapidly changing environments and landscapes (Burroughs 2005; Richerson et al. 2001). Yet, people followed the migrating herds of reindeer as they began to shift their ranges northwards. Much can be said about the differences in ecological inheritance between the four major techno-complexes of the Southern Scandinavian Late Palaeolithic (see Riede 2005b, a, 2007), but here the focus will be on mobility and technology.

The pioneer settlers of Southern Scandinavia, in particular the Hamburgians were specialized reindeer hunters (Bokelmann 1979, 1991). The following of reindeer herds does provide a viable economic base (see discussions in Baales 1996), but puts enormous mobility pressures on the groups practicing it. Petersen & Johansen (1993) have reconstructed the reindeer hunters' settlement pattern as linear, with sites located along the hypothesized reindeer migration routes. Such linearity is decidedly mal-adaptive, especially under conditions of generally low population density (Mandryk 1993; Wobst 1974, 1976). In addition, it is known ethnographically that reindeer herds undergo pronounced fluctuations in terms of size and migration patterns (e.g., Minc & Smith 1989; Stenton 1991) making the maintenance of such a overspecialized economy difficult and making the human groups themselves highly prone to catastrophic demographic fluctuations (David 1973; Minc & Smith 1989). Recent hunter-gatherers in high latitudes solve these problems through applying buffering mechanisms in the form of storage or exchange, but notably there is no evidence for storage in the archaeological record of the Scandinavian Late Glacial (Bratlund 1994) and social storage through trade connections (Halstead & O'Shea 1982) was not possible in this "life without close neighbours" (Åkerlund 2002: 43). Hamburgian hunting economy was not as efficient as that of later periods (Bratlund 1996) and their mobility seems strictly tied to particular landscape types (Arts & Deeben 1987; Tromnau 1975) demonstrating a lack of local landscape knowledge critical for survival in unknown territories (Meltzer 2003). But why then would these groups move into an unknown landscape and not change their behavioural patterns?

The key to understanding the Hamburgian lies with its ancestral culture the Late Magdalenian of northern France and the Low Countries. Not only do they use very similar tools, but they also follow a largely identical economic strategy (Rensink 1995; Schmider 1982, 1987). In terms of mobility and economy, the Hamburgian expansion can be seen as a case of counteractive niche construction, of people trying to maintain an ancestral life style in the face of a changing environment. The notion that the Hamburgian was conservative is supported by an analysis of the lithic material, which compared to later periods is a more complex technology, perfectly fitted to having scarcer and perhaps more distant and varied lithic resources" (Madsen 1992: 128), despite the ready availability of flint in the area (Petersen 1993). Hamburgian technology also lacks high levels of curation (Fig. 11.3) with few

tools being actually used (Narr et al. 1989) and much breakage during blank production (Madsen 1992). Again, such practices seem mal-adaptive and are clear holdovers from ancestral cultural niche behaviour.

Later groups abandon this high-end technology for a coarser, yet more effective approach with a sudden break at the Older Dryas cold period. They also abandon the reindeer-herd following strategy in favour of elk hunting (Andersson et al. 2004; Bokelmann et al. 1983) and later still in favour of a reindeer hunting economy that relies on the use of the bow and arrow as a more efficient hunting weapon, landscape modifications for mass-drives and waterborne transport for both increased hunting success as well as for swift long-distance transport (Bokelmann 1991; Kindgren 1996; Tromnau 1984, 1987). While the economic changes during the warmer phases of the Late Glacial were certainly in part conditioned by climate changes, the comparison between the Hamburgian and the Ahrensburgian in instructive. The latter techno-complex uses a far more extensive range of technological aids in order to keep up with the herds and engaged in exchange in time of need (Kindgren 2002). The Ahrensburgian, as we saw above, were genetically successful and have left both cultural (Bang-Andersen 2003; Kindgren 2002; Schmitt et al. 2006) as well as genetic (see above) descendants in Northern Fennoscandia. Their survival was facilitated by more flexible niche construction behaviour. The hypothesis hardens that prior to the emergence of the Ahrensburgian, a significant bottleneck in biological, ecological and cultural inheritance took place. The tool repertoire changes significantly a number of times as do economic strategies. Hamburgian elements disappear after the Bølling warm period, but it remains unclear how the Federmesser, Bromme and Ahrensburgian cultures are related. An investigation of the micro-evolutionary dynamics of material culture change may assist in evaluating these relationships.

CULTURAL INHERITANCE

All human cultures have extra-somatic, material means of engaging with the environment. In Southern Scandinavia, the four major techno-complexes are defined through their lithics and, in particular, through projectile point morphology. These manufacturing traditions were maintained through teaching and learning as documented at the Bromme site Trollesgave, Denmark (Fischer 1989b, 1990). Refitting and technological analyses allow the identification of different levels of skill within a group as well as actual situations of formal teaching. These on-the-ground archaeological contexts provide the rationale for linking this "palaeontology of the gesture" (Schlanger 1990: 145) with phylogenetic reconstructions of such lineages of social information transmission within populations (Riede 2006).

Based on previous technological analyses (Hartz 1987; Ikinger 1998; Madsen 1992, 1996), a recording system

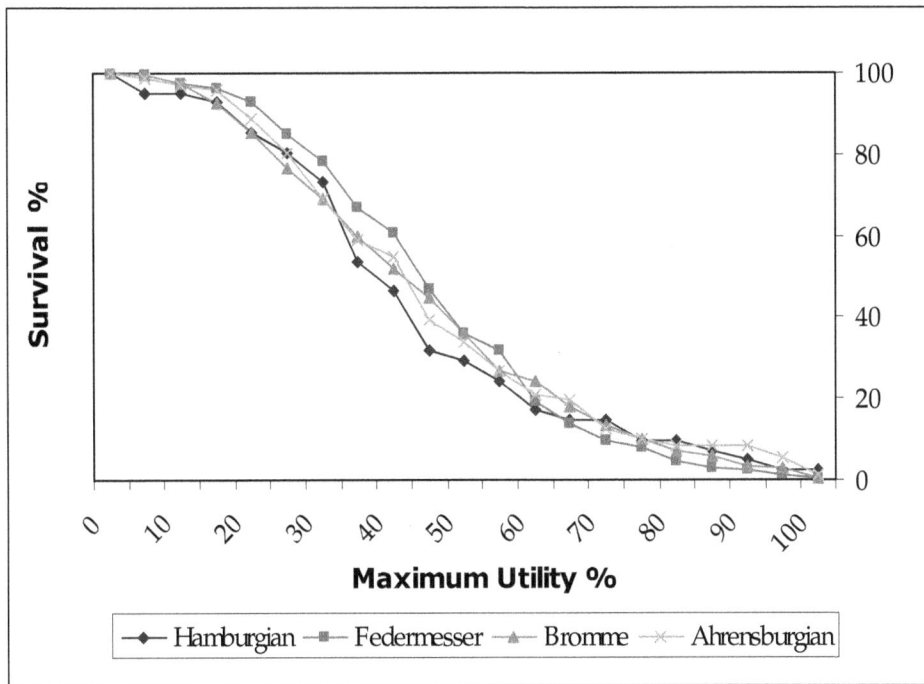

Fig. 11.3. A quantitative analysis of curation in scraping tools of the Late Glacial forager groups, following the method presented by Shott (1993), finds no evidence for higher curation rates in the Hamburgian. The total sample of 1577 specimens measured by the author is divided by cultural group/site context (as indicated by the excavator or curator). In support of the view that Hamburgian tools may have been maladaptive in the Southern Scandinavian context, Kuhn (1994) suggests that the ideal size of scraping tools should be only 1.5 times their length. Remarkable, the Federmesser groups are distinguished by scraping tools (so-called thumb-nail scrapers) that fit this prediction reasonably well

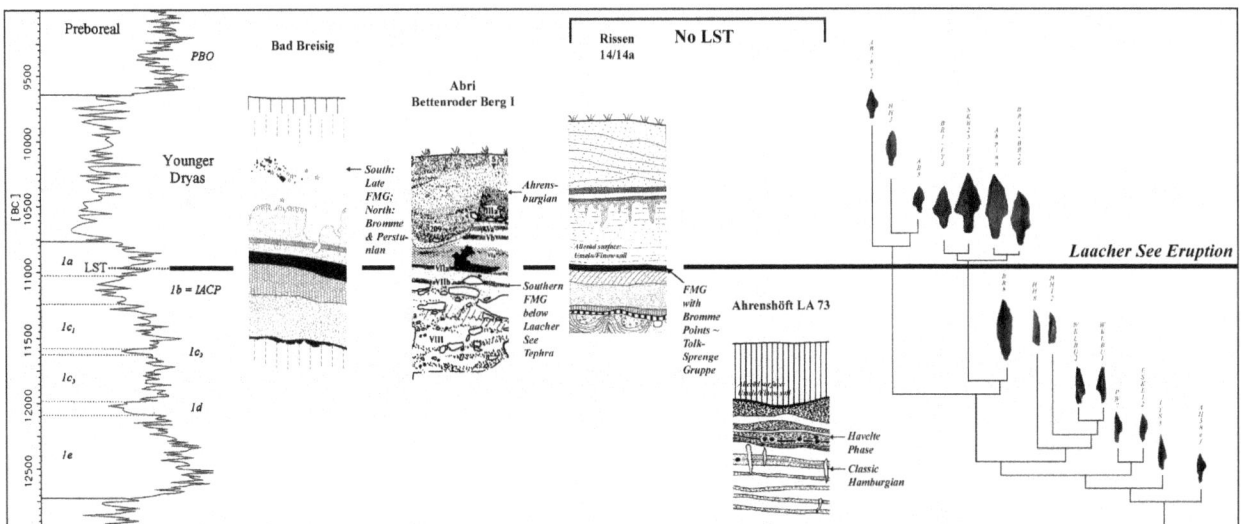

Fig. 11.4. A maximum likelihood tree of 16 Late Glacial projectile point taxa juxtaposed to key stratigraphic sequences from Ahrenshöft, Rissen, Bettenroder Berg and Bad Breisig and the $\delta^{18}O$ temperature proxy data of the GISP2 ice core. LST = Laacher See-tephra

was designed to track the transmission of knapping skills through time in Late Glacial projectile points. Because reticulation cannot be excluded for cultural evolution (Borgerhoff-Mulder *et al.* 2006), network methods (Bandelt *et al.* 1999) were initially used to assess the extent of horizontal transmission within the dataset. Such networks reveal much detail about the data structure and allow interpretations regarding the flow of information along branches (Bryant *et al.* 2005; Lipo 2006; see also Morrison 2005). From these networks, trees can be extracted, and the approach follows the method of O'Brien *et al.* 2001; O'Brien & Lyman 2000, 2002, 2003;

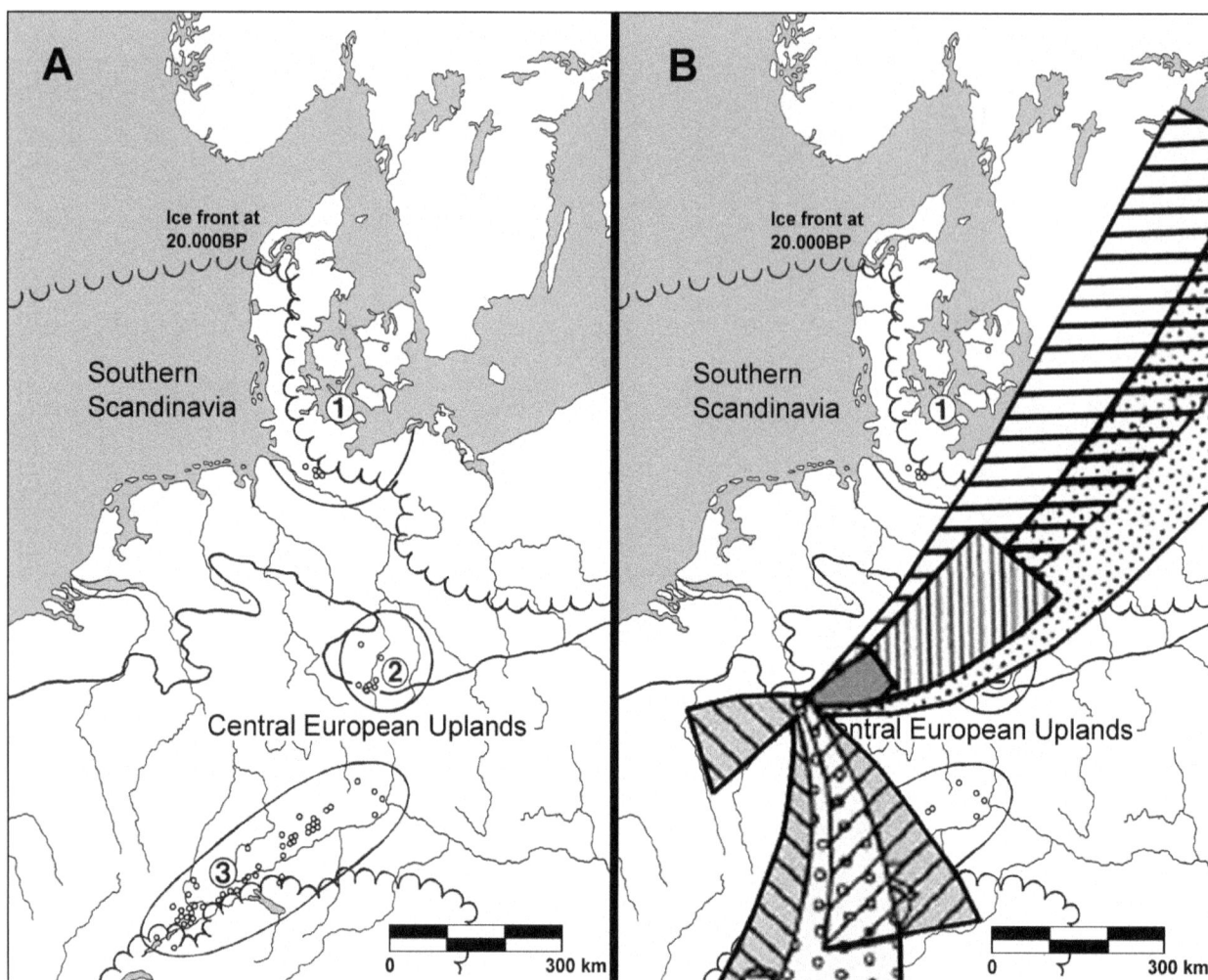

Fig. 11.5. Eriksen's (1996) map of the European Late Glacial, which corresponds to Figure 11.2. The Laacher See-tephra cuts Southern Scandinavia off from the core settlement areas. Interestingly, the Thuringian Basin (2) appears to have been depopulated prior to the Laacher See-eruption, leaving this key area as a Palaeolithic no-man's land after groups moved away from regions affected by ash fall-out. Cultural historical schemes suggest that even in south-central Europe, the demographic fluctuations instigated by the Laacher See-eruption may have led to (largely stylistic) cultural changes

O'Brien *et al.* 2002). Figure 11.4 shows a maximum likelihood tree for the 16 taxa used juxtaposed to climatic and stratigraphic data.

The discovery of a stratigraphic separation between 'classic' Hamburgian and the Havelte phase (Clausen 1998) has prompted some workers to distinguish sub-phases on the higher taxonomic level of the culture (Terberger 1996). The taxa that can be said to belong to a Hamburgian clade are often on isolated branches with high taxonomic homogeneity but low bootstrap values. This is here interpreted as reflecting relatively isolated micro-traditions and perhaps several extinction-coloni-zation cycles. Soltis *et al.* (1995) have shown that even moderate increases in mortality rates can lead to noti-ceable cultural change and induce people to adopt the practices of other groups. The later periods show rather more continuity – arch-backed points and large tanged points are part of the hunting arsenal of peripheral Azilien

groups (Breest 1999; Gerken 2001b, a; Mace 1959) until the catastrophic eruption of the Laacher See volcano (Baales *et al.* 2002; Litt *et al.* 2003) which disrupted traditional communication routes and precipitated cultural change through loss of connectivity among fringe groups (Riede in prep.; Fig. 11.5). Henrich's (2004) model for demographically contingent mal-adaptive loss of techno-logical complexity is particularly appropriate here as a quantitative analysis of the Bromme points (Table 11.1; Fig. 11.6) indicates that these were in fact dart points rather than arrowheads (*contra* Fischer 1989a; Fischer *et al.* 1984). At the Laacher See eruption the slender Federmesser points are lost and with them bow and arrow technology. There is little doubt that this technology is adaptive (Rozoy 1985) and such a loss would stand against adaptationist reconstructions of projectile point evolution. Bromme points become smaller as populations are pushed southwards during the Younger Dryas and come into contact with bow-and-arrow using groups.

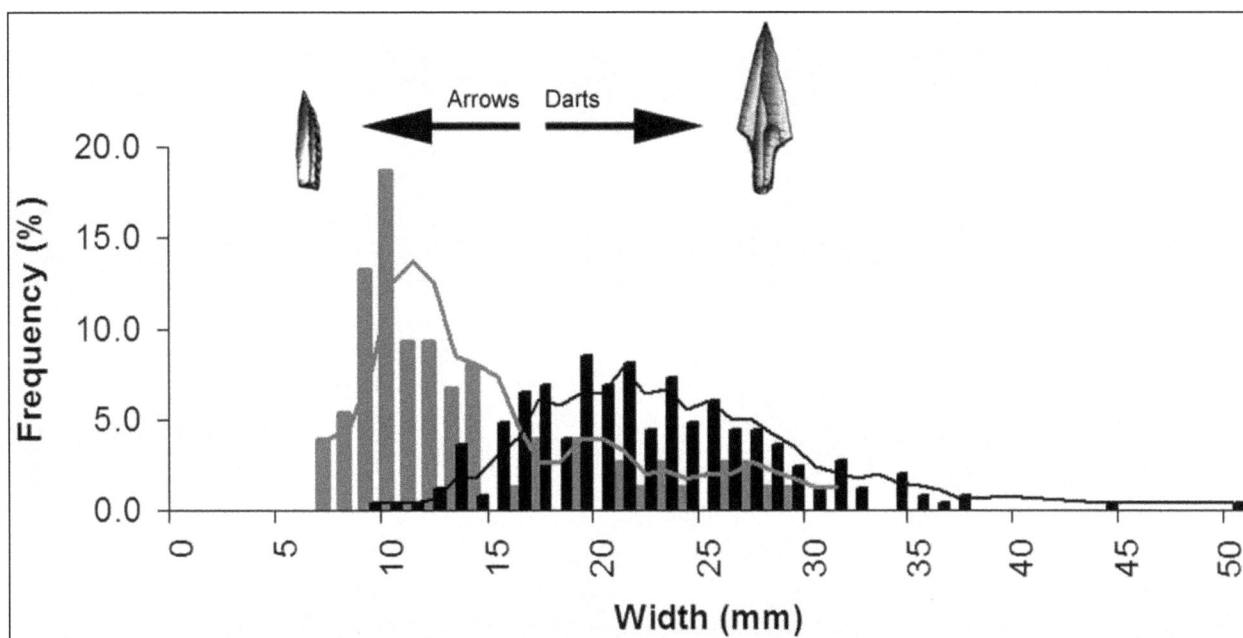

Fig. 11.6. The inconsistency in the classification of Federmesser and Bromme points in Table 1 is resolved when the measurement distributions of the pivotal variable, maximal width, is plotted as a simple histogram. The Federmesser sample is bimodal with the subsidiary mode representing large tanged points found in Federmesser contexts. These indicate the use of a secondary weapon delivery system (the dart and spear-thrower) in addition to the bow-and-arrow on the Northern European Plain prior to the Laacher See-eruption. After the eruption, the bow-and-arrow disappears from the repertoire and is replaced by the exclusive use of darts. Those Bromme specimens identified as arrows either denote a late trend towards arrow tips or a misidentification of these points inherent in the method used.

Tab. 11.1. The classification of the measured Late Palaeolithic projectile points following Shott's (1997) discriminant function analysis (one-trait variant using width) and using measurements taken by the author on complete or nearly complete specimens. Note that arrow shafts are known from Ahrensburgian contexts. The methods ability to correctly classify these projectiles as arrow heads lends confidence that the other assessments are also largely correct (the method is known, however, to underscore dart points)

Cultural Complex	Weapon System	N	%
Federmesser	Arrow	61	81.3
	Dart	14	18.7
Bromme	Arrow	73	29.4
	Dart	175	70.6
Ahrensburgian	Arrow	134	95
	Dart	7	5

CONCLUSION: NICHE CONSTRUCTION LESSONS FOR THE LATE PALAEOLITHIC OF SOUTHERN SCANDINAVIA

This study was designed following the guidelines of the niche construction model. In an attempt to present an integrated Darwinian view on the Late Glacial re-colonization of Southern Scandinavia, genetic, ecological and material cultural data was harnessed. Although it remains challenging to integrate information from deep prehistory with the formal aspects of the niche construction model, it does serve as a useful and open-ended overall framework. The need to incorporate biological inheritance and to relate it explicitly to the ecological and cultural data is stressed.

With respect to the case study presented here, all three lines of enquiry point towards more structured demographic dynamics of the Late Glacial re-colonization process. In particular the earliest period, commonly referred to as the Hamburgian stands out as an ultimately unsuccessful colonization attempt and a significant break in all three tiers of inheritance can be demonstrated. It is likely that such a collapse was followed by the merging of the surviving individuals into more stable populations, but evidence of genetic and cultural input of the earliest groups into later ones is marginal. This process can be seen as counteractive niche construction leading to negative niche construction. Population and cultural continuity can be demonstrated from the Allerød period onwards. However, despite evidence for continuity, considerable cultural change and ecological range shifts took place. It is clear that we cannot sideline climate change in our reconstructions of bio-social change in prehistory. Here it was the stress produced by high-amplitude and high-magnitude environmental change that led to population collapse (Diamond 2005), but equally episodes of unique catastrophic climate upheavals impacted noticeably on the course of evolution (Grattan 2006). Demography acts as an important middle-range link between the different levels of inheritance and the "grand changes" (Price 1991: 189) in artefact morphology during this period may reflect significant demographic fluctuations. With very small populations living at extremely low population densities (Bocquet-Appel *et al.* 2005), prehistoric foragers would have been susceptible to climate change on stochastic grounds and due their relatively limited capacities of buffering niche construction. After all, "even the most adaptable of creatures will experience limits to its tolerance space, outside of which it is unable to behave adaptively" (Laland & Brown 2006: 98). Future work should focus on developing more formal approaches that allow archaeological data to be used in niche construction models.

References

ACHILI, A., C. RENGO, C. MAGRI, V. BATTAGLIA & A. OLIVIERI 2004. The Molecular Dissection of mtDNA Haplogroup H Confirms That the Franco-Cantabrian Glacial Refuge Was a Major Source for the European Gene Pool. *American Journal of Human Genetics* 75: 910-18.

ÅKERLUND, A. 2002. Life without close neighbours. Some reflections on the first peopling of east Central Sweden, in B.V. Eriksen & B. Bratlund (eds.) *Recent studies in the Final Palaeolithic of the European Plain*. 43-48. Højbjerg: Jutland Archaeological Society.

ANDERSSON, M., P. KARSTEN, B. KNARRSTRÖM & M. SVENSSON 2004. *Stone Age Scania. Significant places dug and read by contract archaeology.* Lund: Riksantikvarieämbetet.

ARTS, N. & J. DEEBEN 1987. On the Northwestern Border of Late Magdalenian Territory: Ecology and Archaeology of Early Late Glacial Band Societies in Northwestern Europe, in J.M. Burdukiewicz & M. Kobusiewicz (eds.) *Late Glacial in Central Europe. Culture and Environment.* 5: 25-66. Wroclaw: Prace Komisji Archeologicznej.

BAALES, M. 1996. *Umwelt und Jagdökonomie der Ahrensburger Rentierjäger im Mittelgebirge.* Bonn: Verlag Rudolf Habelt GmbH.

BAALES, M., O. JÖRIS, M. STREET, F. BITTMANN, B. WENINGER & J. WIETHOLD 2002. Impact of the Late Glacial Eruption of the Laacher See Volcano, Central Rhineland, Germany. *Quaternary Research* 58: 273-88.

BANDELT, H.-J., P. FORSTER & A. RÖHL 1999. Median-joining networks for inferring intraspecific phylo-genies. *Molecular Biology and Evolution* 16: 37-48.

BANG-ANDERSEN, S. 2003. Southwest Norway at the Pleistocene/Holocene Transition: Landscape Development, Colonization, Site Types, Settlement Patterns. *Norwegian Archaeological Review* 36: 5-25.

BOCQUET-APPEL, J.-P., P.-Y. DEMARS, L. NOIRET & D. DOBROWSKY 2005. Estimates of Upper Palaeolithic meta-population size in Europe from archaeological data. *Journal of Archaeological Science* 32: 1656-68.

BOKELMANN, K. 1979. Rentierjäger am Gletscherrand in Schleswig-Holstein? Ein Diskussionsbeitrag zur Erforschung der Hamburger Kultur. *Offa* 36: 12-22.

BOKELMANN, K. 1991. Some new thoughts on old data on humans and reindeer in the Ahrensburg tunnel valley in Schleswig-Holstein, Germany, in R.N.E. Barton, A.J. Roberts & D.A. Roe (eds.) *The Late-Glacial in North-West Europe.* CBA Report. 77: 72-81. Oxford: CBA.

BOKELMANN, K., D. HEINRICH & B. MENKE 1983. Fundplätze des Spätglazials am Hainholz-Esinger Moor, Kreis Pinneberg. *Offa* 40: 199-239.

BORGERHOFF-MULDER, M., C.L. NUNN & M.C. TOWNER 2006. Cultural Macroevolution and the Transmission of Traits. *Evolutionary Anthropology* 15: 52-64.

BRATLUND, B. 1994. A survey of subsistence and settlement pattern of the Hamburgian culture in Schleswig-Holstein. *Jahrbuch des Römisch-Germanischen Zentralmuseums Mainz* 41: 59-93.

BRATLUND, B. 1994. 1996. Hunting Strategies in the Late Glacial of Northern Europe: A Survey of the Faunal Evidence. *Journal of World Prehistory* 10: 1-48.

BREEST, K. 1999. Der spätpaläolithische Oberflächenfundplatz mit Rücken- und Bromme-Spitzen bei Dohnsen-Bratzloh, Ldkr. Celle (Niedersachsen), in E. Cziesla, T. Kersting & S. Pratsch (eds.) *Den Bogen spannen...Festschrift für Bernhard Gramsch.* 1: 67-75. Weissbach: Beier & Beran.

BRYANT, D., F. FILIMON & R.D. GRAY 2005. Untangling our Past: Languages, Trees, Splits and Networks, in R. Mace, C.J. Holden & S.J. Shennan (eds.) *The Evolution of Cultural Diversity. A Phylogenetic Approach.* 67-83. London: UCL Press.

BURROUGHS, W.J. 2005. *Climate Change in Prehistory. The End of the Reign of Chaos.* Cambridge: Cambridge University Press.

CAVALLI-SFORZA, L.L. & M. FELDMAN 2003. The application of molecular genetic approaches to the study of human evolution. *Nature Genetics* 33: 266-75.

CLAUSEN, I. 1998. Neue Untersuchungen an späteiszeitlichen Fundplätzen der Hamburger Kultur bei Ahrenshöft, Kr. Nordfriesland. Ein Vorbericht. *Archäologische Nachrichten aus Schleswig-Holstein* 8: 8-49.

COLLARD, M. & S.J. SHENNAN 2000. Processes of Culture Change in Prehistory: a Case Study from the European Neolithic, in C. Renfrew & K.V. Boyle (eds.) *Archaeogenetics: DNA and the population prehistory of Europe*. 89-97. Cambridge: McDonald Institute for Archaeological Research.

COLLARD, M. & J.J. TEHRANI 2005. Phylogenesis versus Ethnogenesis in Turkmen Cultural Evolution, in R. Mace, C.J. Holden & S.J. Shennan (eds.) *The Evolution of Cultural Diversity. A Phylogenetic Approach*. 109-32. London: UCL Press.

CROES, D., K.M. KELLY & M. COLLARD 2005. Cultural historical context of Qwu?gwes (Puget Sound, USA): a preliminary investigation. *Journal of Wetland Archaeology* 5: 141-54.

DAVID, N. 1973. On upper palaeolithic society, ecology, and technological change: the Noaillian case, in C. Renfrew (ed.) *The Explanation of Culture Change. Models in Prehistory*. 277-303. London: Duckworth.

DAY, R.L., K.N. LALAND, J. ODLING-SMEE & M.W. FELDMAN 2003. Rethinking Adaptation: The Niche Construction Perspective. *Perspectives in Biology and Medicine* 46: 80-95.

DIAMOND, J.M. 2005. *Collapse. How Societies Choose to Fail or Survive*. London: Penguin Books.

ERIKSEN, B.V. 1996. Regional Variation in Late Pleistocene Subsistence Strategies. Southern Scandinavian Reindeer Hunters in a European Context, in L. Larsson (ed.) *The Earliest Settlement of Scandinavia and its relationship with neighbouring areas*. Acta Archaeologica Lundensia Series IN 8°, No. 24. 7-22. Stockholm: Almqvist & Wicksell.

FISCHER, A. 1989a. Hunting with Flint-Tipped Arrows: Results and Experiences from Experiments, in C. Bonsall (ed.) *The Mesolithic in Europe*. 29-39. Edinburgh: John Donald.

FISCHER, A. 1989b. A Late Palaeolithic "School" of Flint-Knapping at Trollesgave, Denmark. Results from Refitting. *Acta Archaeologica* 60: 33-49.

FISCHER, A. 1990. On Being a Pupil of a Flintknapper of 11,000 Years Ago. A preliminary analysis of settlement organization and flint technology based on conjoined flint artefacts from the Trollesgave site, in E. Cziesla, S. Eickhoff, N. Arts & D. Winter (eds.) *The Big Puzzle: International Symposium on Refitting Stone Artefacts, Monrepos, 1987*. 447-64. Bonn: Holos.

FISCHER, A., P.V. HANSEN & P. RASMUSSEN 1984. Macro and Micro Wear Traces on Lithic Projectile Points. Experimental Results and Prehistoric Examples. *Journal of Danish Archaeology* 3: 19-46.

FITZHUGH, B. 2001. Risk and Invention in Human Technological Evolution. *Journal of Anthropological Archaeology* 20: 125-67.

FORSTER, P. 2004. Ice Ages and the mitochondrial DNA chronology of human dispersals: a review. *Philosophical Transactions of the Royal Society of London, Series B* 359: 255-64.

FUGLESTVEDT, I. 2005. Contact and communication in Northern Europe 10.200-9.000/8.500 BP – a phenomenological approach to the connection between technology, skill and landscape, in H. Knutsson (ed.) *Pioneer settlement and colonization processes in the Barents region*. 1: 79-96. Vuollerim: Vuollerim6000år.

GERKEN, K. 2001a. Westertimke 69 – eine Jagdstation der Federmesser-Gruppen, in B. Gehlen, M. Heinen & A. Tillmann (eds.) *Zeit-Räume. Gedenkschrift für Wolfgang Taute*. Archäologische Berichte 14. 363-80. Bonn: Verlag Rudolf Habelt GmbH.

GERKEN, K. 2001b. *Studien zur jung- und spätpaläolithischen sowie mesolithischen Besiedlung im Gebiet zwischen Wümme und Oste*. Archäologische Berichte des Landkreises Rotenburg (Wümme). 9. Oldenburg: Isensee Verlag.

GOLDSTEIN, D.B. & L. CHIKHI 2002. Human Migrations and Population Structure: What We Know and Why it Matters. *Annual Review of Genomics and Human Genetics* 3: 129-52.

GRATTAN, J. 2006. Aspects of Armageddon: An exploration of the role of volcanic eruptions in human history and civilization. *Quaternary International* 151: 10-18.

HALSTEAD, P. & J. O'SHEA 1982. A friend in need is a friend indeed: social storage and the origins of social ranking, in C. Renfrew & S.J. Shennan (eds.) *Ranking, Resource and Exchange*. 92-99. Cambridge: Cambridge University Press.

HARDESTY, D.L. 1972. The Human Ecological Niche. *American Anthropologist* 74: 458-66.

HARTZ, S. 1987. Neue spätpaläolithische Fundplätze bei Ahrenshöft, Kreis Nordfriesland. *Offa* 44: 5-52.

HENRICH, J. 2004. Demography and Cultural Evolution: How Adaptive Cultural Processes Can Produce Maladaptive Losses – the Tasmanian Case. *American Antiquity* 69: 197-214.

IKINGER, E.-M. 1998. *Der endeiszeitliche Rückenspitzen-Kreis Mittleuropas*. GeoArchaeoRhein, Nr.1. Münster: LIT.

JOBLING, M.A., M.E. HURLES & C. TYLER-SMITH 2004. *Human Evolutionary Genetics. Origins, Peoples & Diseases*. New York, N.Y.: Taylor & Francis.

JORDAN, P. & T. MACE 2006. Tracking Culture-Historical Lineages: Can "Descent with Modification" be Linked to "Association by Descent"?, in C.P. Lipo, M.J. O'Brien, M. Collard & S.J. Shennan (eds.) *Mapping out Ancestors. Phylogenetic Approaches in Anthropology and Prehistory*. 149-68. New Brunswick: Aldine Transaction.

KARLSSON, A.O., T. WALLERSTROM, A. GOTHERSTROM & G. HOLMLUND 2006. Y-chromosome diversity in Sweden – A long-time perspective. *European Journal Human Genetics* 14: 963-70.

KINDGREN, H. 1996. Reindeer or seals? Some Late Palaeolithic sites in central Bohuslän, in L. Larsson (ed.) *The Earliest Settlement of Scandinavia and Its Relationship with Neighbouring Areas*. 191-205. Stockholm: Almqvist & Wiksell.

KINDGREN, H. 2002. Tosskärr. Stenkyrka 94 revisited, in B.V. Eriksen & B. Bratlund (eds.) *Recent studies in the Final Palaeolithic of the European plain*. 49-60. Højbjerg: Jutland Archaeological Society.

KUHN, S.L. 1994. A Formal Approach to the Design and Assembly of Mobile Toolkits. *American Antiquity* 59: 426-42.

LALAND, K.N. & G.R. BROWN 2006. Niche Construction, Human Behavior, and the Adaptive-Lag Hypothesis. *Evolutionary Anthropology* 15: 95-104.

LALAND, K.N. & K. STERELNY 2006. Perspective: Seven Reasons (Not) to Neglect Niche Construction. *Evolution* 60: 1751-62.

LALAND, K.N., F.J. ODLING-SMEE & M.W. FELD-MAN 1999. Evolutionary consequences of niche construction and their implications for ecology. *Proceedings of the National Academy of Sciences of the United States of America* 96: 10242-47.

LALAND, K.N., F.J. ODLING-SMEE & M.W. FELD-MAN 2000. Niche construction, biological evolution, and cultural change. *Behavioural and Brain Sciences* 23: 131-75.

LALAND, K.N., F.J. ODLING-SMEE & M.W. FELD-MAN 2001. Cultural niche construction and human evolution. *Journal of Evolutionary Biology* 14: 22-33.

LALAND, K.N., F.J. ODLING-SMEE & M.W. FELD-MAN 2004. Causing a commotion. *Nature* 429: 609.

LEWONTIN, R.C. 2000. *The Triple Helix. Gene, Organism and Environment*. Cambridge, MA: Harvard University Press.

LIPO, C.P. 2006. The Resolution of Cultural Phylogenies Using Graphs, in C.P. Lipo, M.J. O'Brien, M. Collard & S.J. Shennan (eds.) *Mapping Our Ancestors. Phylogenetic Approaches in Anthropology and Prehistory*. 89-108. New Brunswick, N.J.: AldineTransaction.

LITT, T., H.-U. SCHMINCKE & B. KROMER 2003. Environmental response to climatic and volcanic events in central Europe during the Weichselian Lateglacial. *Quaternary Science Reviews* 22: 7-32.

LOOGVÄLI, E.-L., U. ROOSTALU, B.A. MALYAR-CHUCK, M.V. DERENKO, T. KIVISILD, E. METSPALU & K. TAMBETS 2004. Disuniting Uniformity: A Pied Cladistic Canvas of mtDNA Haplogroup H in Eurasia. *Molecular Biology and Evolution* 21: 2012-21.

MACE, A. 1959. An Upper Palaeolithic Open-site at Hengistbury Head, Christchurch, Hants. *Proceedings of the Prehistoric Society* 25: 233-59.

MADSEN, B. 1992. Hamburgkulturens flintteknologi i Jels (The Hamburgian Flint Technology at Jels), in J. Holm & F. Rieck (eds.) *Istidsjægere ved Jelssøerne*. 5: 93-132. Haderslev: Skrifter fra Museumsrådet for Sønderjyllands Amt.

MADSEN, B. 1996. Late Palaeolithic cultures of south Scandinavia: tools, traditions and technology, in L. Larsson (ed.) *The Earliest Settlement of Scandinavia and Its Relationship with Neighbouring Areas*. Acta Archaeologica Lundensia Series IN 8°. No.24: 61-73. Stockholm: Almqvist & Wiksell.

MANDRYK, C.A.S. 1993. Hunter-Gatherer Social Costs and the Nonviability of Submarginal Environments. *Journal of Anthropological Research* 49: 39-71.

MELTZER, D. 2003. Lessons in landscape learning, in M. Rockman & J. Steele (eds.) *Colonization of Unfamiliar Landscapes: The archaeology of adaptation*. 222-41. London: Routledge.

MINC, L.D. & K.P. SMITH 1989. The spirit of survival, in P. Halstead & J. O'Shea (eds.) *Bad Year Economics: Cultural responses to risk and uncertainty*. 8-39. Cambridge: Cambridge University Press.

MORRISON, D.A. 2005. Networks in phylogenetic analysis: new tools for population biology. *International Journal of Parasitology* 35: 567-82.

NARR, K.J., M. HAMÖLLER & X. NAVARRO HARRIS 1989. Gebrauchsspuren an Artefakten der Hamburger Kultur von Stellmoor, Kreis Stormarn. *Offa* 46: 5-16.

NEWELL, R.R., T.S. CONSTANDSE-WESTERMANN & C. MEIKLEJOHN 1979. The skeletal remains of Mesolithic man in western Europe: an evaluative catalogue. *Journal of Human Evolution* 8: 1-228.

O'BRIEN, M.J. & R.L. LYMAN 2000. Evolutionary Archaeology. Reconstructing and Explaining Historical Lineages, in M.B. Schiffer (ed.) *Social Theory in Archaeology*. 126-42. Salt Lake City, UT.: University of Utah Press.

O'BRIEN, M.J. & R.L. LYMAN 2002. Evolutionary Archaeology: Current Status and Future Prospects. *Evolutionary Anthropology* 11: 26-36.

O'BRIEN, M.J. & R.L. LYMAN 2003. Resolving phylogeny: Evolutionary archaeology's fundamental issue, in T.L. VanPool & C.S. VanPool (eds.) *Essential Tensions in Archaeological Method and Theory*. 115-35. Salt Lake City, UT.: University of Utah Press.

O'BRIEN, M.J., J. DARWENT & R.L. LYMAN 2001. Cladistics is useful for reconstructing archaeological phylogenies: Paleoindian points from the southeastern United States. *Journal of Archaeological Science* 28: 1115-36.

O'BRIEN, M.J., R.L. LYMAN, Y. SAAB, E. SAAB, J. J. DARWENT & D.S. GLOVER 2002. Two issues in archaeological phylogenetics: Taxon construction and outgroup selection. *Journal of Theoretical Biology* 215: 133-50.

ODLING-SMEE, F.J. 2006. How Niche Construction Contributes to Human Gene-Culture Coevolution, in J.C.K. Wells, S. Strickland & K.N. Laland (eds.) *Social Information Transmission and Human Biology.* 39-58. London: CRC Press.

ODLING-SMEE, F.J., K.N. LALAND & M.W. FELDMAN 1996. Niche Construction. *American Naturalist* 147: 641-48.

ODLING-SMEE, F.J. 2003. *Niche Construction. The Neglected Process in Evolution.* Princeton: Princeton University Press.

OSWALT, W.H. 1976. *An Anthropological Analysis of Food-Getting Technology.* New York, N.Y.: John Wiley & Sons.

PASSARINO, G., G.L. CAVALLERI, A.A. LIN, L.L. CAVALLI-SFORZA, A.-L. BØRRESEN-DALE & P.A. UNDERHILL 2002. Different genetic components in the Norwegian population revealed by the analysis of mtDNA and Y chromosome polymorphisms. *European Journal Human Genetics* 10: 521-29.

PEREIRA, L., M. RICHARDS, A. GOIOS, A. ALONSO, C. ALBARRAN, O. GARCIA, D.M. BEHAR, M. GOLGE, J. HATINA, L. AL-GAZALI, D.G. BRADLEY, V. MACAULAY & A. AMORIM 2005. High-resolution mtDNA evidence for the late-glacial resettlement of Europe from an Iberian refugium. *Genome Research* 15: 19-24.

PETERSEN, P.V. 1993. *Flint fra Danmarks Oltid.* Copenhagen: Høst & Søn.

PETERSEN, P.V. & L. JOHANSEN 1993. Sølbjerg I – An Ahrensburgian Site on a Reindeer Migration Route through Eastern Denmark. *Journal of Danish Archaeology* 10: 20-37.

PRICE, T.D. 1991. The View from Europe: Concepts and Questions about Terminal Pleistocene Societies, in T.D. Dillehay & D. Meltzer (eds.) *First Americans: Search and Research.* 185-208. Boca Raton, FL.: CRC Press.

RENSINK, E. 1995. On magdalenian mobility and land use in north-west Europe. Some methodological considerations. *Archaeological Dialogues* 2: 85-119.

RICHERSON, P.J. & R. BOYD 2005. *Not by genes alone: how culture transformed human evolution.* Chicago: University of Chicago Press.

RICHERSON, P.J., R. BOYD & R.L. BETTINGER 2001. Was Agriculture Impossible During the Pleistocene But Mandatory During the Holocene? A Climate Change Hypothesis. *American Antiquity* 66: 387-411.

RIEDE, F. 2005a. To Boldly Go Where No (Hu-)Man Has Gone Before. Some Thoughts on The Pioneer Colonisations of Pristine Landscapes. *Archaeological Review from Cambridge* 20: 20-38.

RIEDE, F. 2005b. Darwin vs. Bourdieu. Celebrity Deathmatch or Postprocessual Myth? Prolegomenon for the Reconciliation of Agentive-Interpretative and Ecological-Evolutionary Archaeology, in H. Cobb, S. Price, F. Coward & L. Grimshaw (eds.) *Investigating Prehistoric Hunter-Gatherer Identities: Case Studies from Palaeolithic and Mesolithic Europe.* BAR. 1411: 45-64. Oxford: Oxbow.

RIEDE, F. 2006. Chaîne Opèratoire – Chaîne Evolutionaire. Putting Technological Sequences in Evolutionary Context. *Archaeological Review from Cambridge* 21: 50-75.

RIEDE, F. 2007. Stretched thin, like butter on too much bread…Some thoughts about journeying in the unfamiliar landscapes of late Palaeolithic Southern Scandinavia, in R. Johnson & V. Cummings (eds.) *Prehistoric Journeys.* not yet published. Oxford: Oxbow.

RIEDE, F. in prep. The Eruption of the Laacher See Volcano and the Origin of the Bromme and Perstunian Cultures in Northern Europe. *Lund Archaeological Review*

ROZOY, J.-G. 1985. The Revolution of the Bowmen in Europe, in C. Bonsall (ed.) *The Mesolithic in Europe.* 13-28. Edinburgh: John Donald.

RUST, A. 1942. Über die endglaciale Kulturentwicklung im rechtselbischen Nordwesteuropa, unter Berücksichtigung der geologischen und siedlungsarchäologischen Verhältnisse. *Offa* 6/7: 52-75.

SAJANTILA, A. & S. PÄÄBO 1995. Language replacement in Scandinavia. *Nature Genetics* 11: 359-60.

SAJANTILA, A., A.H. SALEM, P. SAVOLAINEN, K. BAUER, C. GIERIG & S. PÄÄBO 1996. Paternal and maternal DNA lineages reveal a bottleneck in the founding of the Finnish population. *Proceedings of the National Academy of Sciences of the United States of America* 93: 12035-39.

SCHLANGER, N. 1990. Technique as Human Action: Two Perspectives. *Archaeological Review from Cambridge* 9: 18-26.

SCHMIDER, B. 1982. The Magdalenian Culture of the Paris River-Basin and Its Relationship with the Nordic Cultures of the Late Old Stone Age. *World Archaeology* 14: 259-69.

SCHMIDER, B. 1987. Environment and Culture in the Seine Basin during the Late Glacial Period, in J.M. Burdukiewicz & M. Kobusiewicz (eds.) *Late Glacial in Central Europe. Culture and Environment.* 5: 11-24. Wroclaw: Prace Komisji Archeologicznej.

SCHMITT, L., S. LARSSON, C. SCHRUM, I. ALEKSEEVA, M. TOMCZAK & K. SVEDHAGE 2006. 'Why They Came'; The Colonization of the Coast of Western Sweden and its Environmental Context at the End of the Last Glaciation. *Oxford Journal of Archaeology* 25: 1-28.

SHOTT, M.J. 1993. An Exegesis of the Curation Concept. *Journal of Anthropological Research* 52: 259-80.

SOLTIS, J., P. BOYD & P.J. RICHERSON 1995. Can Group-functional behaviors evolve by cultural group selection? An empirical test. *Current Anthropology* 63: 473-94.

STENTON, D.R. 1991. Caribou Population Dynamics and Thule Culture Adaptations on Southern Baffin Island, N.W.T. *Arctic Anthropology* 28: 15-43.

STERELNY, K. 2006. Memes Revisited. *The British Journal for the Philosophy of Science* 57: 145-65.

SYKES, B. 2000. Human Diversity in Europe and Beyond: From Blood Groups to Genes, in C. Renfrew & K.V. Boyle (eds.) *Archaeogenetics: DNA and the population prehistory of Europe*. 23-28. Cambridge: McDonald Institute for Archaeological Research.

TAMBETS, K., S. ROOTSI, T. KIVISILD, H. HELP, P. SERK, E.L. LOOGVALI, H.V. TOLK, M. REIDLA, E. METSPALU, L. PLISS, O. BALANOVSKY, A. PSHENICHNOV, E. BALANOVSKA, M. GUBINA, S. ZHADANOV, L. OSIPOVA, L. DAMBA, M. VOEVODA, I. KUTUEV, M. BERMISHEVA, E. KHUSNUTDINOVA, V. GUSAR, E. GRECHA-NINA, J. PARIK, E. PENNARUN, C. RICHARD, A. CHAVENTRE, J.P. MOISAN, L. BARAC, M. PERICIC, P. RUDAN, R. TERZIC, I. MIKEREZI, A. KRUMINA, V. BAUMANIS, S. KOZIEL, O. RICKARDS, G.F. DE STEFANO, N. ANAGNOU, K.I. PAPPA, E. MICHALODIMITRAKIS, V. FERAK, S. FUREDI, R. KOMEL, L. BECKMAN & R. VILLEMS 2004. The western and eastern roots of the Saami – the story of genetic "outliers" told by mitochondrial DNA and Y chromosomes. *American Journal of Human Genetics* 74: 661-82.

TERBERGER, T. 1996. The early settlement of North-East Germany (Mecklenburg-Vorpommern), in L. Larsson (ed.) *The Earliest Settlement of Scandinavia and its relationship with neighbouring areas*. Acta Archaeologica Lundensia. 111-22. Stockholm: Almquist & Wiksell.

TROMNAU, G. 1975. Die jungpaläolithischen Fund-plätze im Stellmoorer Tunneltal im Überblick. *Hammaburg N.F.* 2: 9-20.

TROMNAU, G. 1984. Rentierjagd während des Spät-paläolithikums von Booten aus? *Hammaburg N.F.* 6: 29-38.

TROMNAU, G. 1987. Late Palaeolithic Reindeer-hunting and the use of boats, in J.M. Burdukiewicz & M. Kobusiewicz (eds.) *Late Glacial in Central Europe. Culture and Environment*. 5: 95-106. Wroclaw: Prace Komisji Archeologicznej.

VAN DER LEEUW, S. 1989. Risk, perception, innovation, in S. van der Leeuw & R. Torrence (eds.) *What's New? A closer look at the process of innovation*. 301-29. London: Unwin Hyman.

WOBST, M. 1974. Boundary conditions for Paleolithic social systems: a simulation approach. *American Antiquity* 39: 147-78.

WOBST, M. 1976. Locational Relationships in Paleolithic Society. *Journal of Human Evolution* 5: 49-58.

ACHEULEAN BIFACE REFINEMENT IN THE HUNSGI-BAICHBAL VALLEY, KARNATAKA, INDIA

C. SHIPTON, M. PETRAGLIA

Leverhulme Centre for Human Evolutionary Studies, University of Cambridge

K. PADDAYYA †

Department of Archaeology, Deccan College Post-Graduate and Research Institute

Abstract: Despite its characterization as a period of homogeneity evidence suggests that there is considerable variation between Acheulean assemblages. This study looks at variation in biface refinement and technology across nine sites from a single valley in South India. Results show consistent site-wise patterns in biface thickness to breadth ratio and biface weight, and flake scar size and flake scar width to length ratio. Two explanations seem to be consistent with these patterns: differing levels of biface rejuvenation between assemblages or different socio-cognitive capabilities between the hominins responsible for manufacturing the different assemblages.

Keywords: Achelean biface- variation- patterns

Résumé: Malgré la caractérisation d'une époque homogène les preuves suggèrent qu'il y a un changement considérable entre les assemblages Acheuléens. Cette étude examine la variation en raffinage et technologie des bifaces au travers de neuf sites d'une seule vallée dans le sud de l'Inde. Les résultats montrent que il y a un lien entre l'épaisseur des bifaces et la proportion largeur et poids, la taille des éclats et la proportion largeur et longueur des éclats de taille. Deux explications semblent compatibles avec ces schémas: une différence de niveaux de ravivage des bifaces entre les assemblages ou différentes capacités socio-cognitives entre les hominiens responsables de la manufacture des différents assemblages.

Mots clés: acheuléens bifaces – variation – schéma

INTRODUCTION

Much has been made of the homogeneity of the Acheulean, as the same basic tool types are manufactured for well over a million years in East Africa, while they are also found as far a-field as North Wales and South India. Elsewhere we have argued that this homogeneity is the product of a propensity for imitation which characterised the interactions of hominins from the very beginning of the Acheulean, and possibly long before (Petraglia *et al.* 2005; Shipton *et al.* in press). Here we focus on variation within the Acheulean, in order to determine if there is more than meets the eye in this apparent period of stasis in human history. The Hunsgi-Baichbal basin in Karnataka State, India, has been targeted for this analysis because it contains an abundance of Acheulean localities of widely varying ages.

THE ACHEULEAN PHENOMENON

The Acheulean is a technocomplex which first appears in the East African Great Rift Valley around 1.7 million years ago (hereafter mya) in the early Pleistocene (Roche & Kibunjia 1994; Dominguez-Rodrigo *et al.* 2001). Its diagnostic lithics are handaxes: tear-drop shaped bifaces; and cleavers: bifaces with broad bits as their tips.

Experimental, archaeological and micro-wear studies have shown that the principal function of handaxes was in all likelihood butchery (Jones 1980; Mitchell 1997; Pitts & Roberts 1998; Roberts & Parfitt 1999), although they may have been used on vegetable matter as well (Keeley

1980; Dominguez-Rodrigo *et al.* 2001). Their pointed tips are good for piercing skin and their long cutting edges, which extend around much of the perimeter, are good for slicing flesh. The elaborate nature of the handaxe suggests that it was not merely designed to scrape scraps of flesh off bone, but to butcher an entire carcass. The function of cleavers is more elusive, as their name suggests they may also have been involved in butchery. Experimental evidence has shown that towards the latter stages of butchery there are processes which handaxes are not well suited for, such as removing limbs and breaking open long bones (Pitts and Roberts 1998). In contrast to handaxes, cleavers have been neglected from most experimental and microwear studies, possibly due to their virtual absence from the European record and the failure of some Acheulean archaeologists to recognise their existence.

The Acheulean may have been born as the most successful of a variety of developed Oldowan industries in the early Pleistocene of East Africa. The earliest Acheulean bifaces tend to be heavy, thick and relatively unstandardised (Roe 1971). The oldest non-African Acheulean site, Ubeidya, occurs in the Jordan Valley at 1.4 mya (Goren-Inbar & Saragusti 1996). Traces of this initial dispersal of the Acheulean have also been found in India, where the site of Isampur Quarry has been preliminarily dated to 1.2 mya (Blackwell *et al.* 2001). Both the Jordan valley, which is the northern extension of the Great Rift Valley, and peninsula India have very similar climates to East Africa, being hot and dry, with highly seasonal rainfall. This factor may go some way to explaining their early occupation by Acheulean hominins.

Around 900 thousand years ago (hereafter kya) a potential new facies of the Acheulean appears in East Africa at Olorgessailie (Isaac 1977; Bye *et al.* 1987); Olduvai Bed IV (Leakey & Roe 1994); Kilombe and Kariandusi (Gowlett & Crompton 1994). Acheulean artefacts appear for the first time in north-west Africa around 900 kya (Raynal & Texier 1989; Raynal *et al.* 1995) and an Acheulean industry again occurs in the Jordan Valley at the site of Gesher Benot Ya'qov. Comparisons of Gesher Benot Ya'qov have shown the assemblage to have African rather than local affinities, suggesting a second wave of dispersal of the Acheulean from Africa (Goren-Inbar & Saragusti 1996).

After 600 kya it is possible to discern a further facies of the Acheulean (Wynn 2002), associated with the species *Homo heidelbergensis*. For the first time the Acheulean is found in Europe during this period, even as far north as Britain (Gamble 1995; Pitts & Roberts 1998). Indeed from this time the frequency of hominin fossil sites and archaeological sites in both Europe and Africa rises dramatically (Roebrooks 1994; Clark 2001; McNabb 2005). This Late Acheulean is characterised by highly symmetrical artefacts (Wynn 2002); widespread use of prepared core techniques (Debono & Goren-Inbar 2001; Clark 1994; White & Ashton 2003); the use of soft-hammers (Pitts & Roberts 1998); and the earliest tentative evidence for symbolically imbued bifaces (Oakley 1981; Wymer 1982; Halstead 1982; Carbonell *et al* 2003).

There is thus considerable evidence for an increase in the sophistication of the Acheulean over time. Previous quantitative studies have shown significant variation in the symmetry and regularity of bifaces between assemblages of different ages (Saragusti et al 1998; Saragusti *et al.* 2005). This study will focus on variation within a single valley where patterns of raw material usage were similar and where the environment is thought to have been relatively stable throughout the hominin occupation.

THE HUNSGI-BAICHBAL VALLEY

The Hunsgi-Baichbal Valley is located in the physiographic region of the Deccan Plateau and in the political district of Gulbarga, Karnataka, roughly in the centre of peninsular India (Paddayya 1982). The basin forms the headwater zone of a minor left-bank tributary of the River Krishna, called the Hunsgi stream. The basin is of Tertiary age covering an area of c 500 km^2 and consists of the two valleys of the Hunsgi and Baichbal rivers, separated by a narrow remnant of a shale-limestone plateau. The basin is flanked on its western and northern sides by a shale-limestone plateau, capped by volcanic basalts of the Deccan Trap, and on its eastern and southern sides by low hills of Dharwar schist and granite. The basin floor slopes gently from west to east (from 480 to 420 m), while the surrounding uplands and hills range in elevation from 20 to 60 m above the basin floor. Seep springs emanate from the junctions of sedimentary rocks and the underlying impervious Archaean formations, forming shallow perennial streams. Thick travertine deposits indicate that the seep springs were present in the Middle Pleistocene (Szabo *et al.* 1990). The valleys are located in a semi-arid tract of peninsular India, with a mean annual precipitation of 600 mm largely falling during the July to September monsoon season.

The enclosed, amphitheatre-like form of the basin, the gently undulating nature of its floor, the low heights of the surrounding hills and tablelands, the availability of perennial water sources, the occurrence of various rocks suitable for tool-making and the availability of a variety of plant and animal foods, promoted continuous hominin occupation of the basin from the Lower Paleolithic onwards. The intensive survey undertaken by Paddayya from 1974 to 1986 led to the discovery of over 200 Acheulean localities, giving the basin one of the densest concentrations of Acheulean sites in India. The sites have been studied from a settlement system perspective (Paddayya 1982; 2001) and from a taphonomic perspective (Paddayya and Petraglia 1995; Jhaldiyal 1997). The absolute dates for the Acheulean in this basin range from 280 kya for the site of Teggihalli II (Szabo *et al.* 1990), to 1.2 mya for the site of Isampur Quarry (Paddayya *et al.* 2002). One of the most unusual features of these sites in an Indian context is the preference for the use of limestone for tool-making. The limestone in the basin weathers in tabular slabs of varying thickness and is highly siliceous, making it ideal for biface manufacture.

From the Acheulean sites of the basin, bifaces from nine sites were selected for the present study:

1. Hunsgi V (Hunsgi valley), a very large excavated site yielding around 150 bifaces (Paddayya 1977).

2. Hunsgi VI (Hunsgi valley), an excavated site containing bifaces on limestone flakes (Paddayya 1979).

3. Yediyapur IV (Baichbal valley) a surface site containing two giant bifaces among other smaller specimens (Paddayya 1987).

4. Yediyapur VI (Baichbal valley) an excavated site where bifaces were produced on a variety of raw materials including quartzite flakes (Paddayya 1987).

5. Mudnur VIII (Baichbal valley), a surface site possibly a cache, where several large bifaces were found with no other cultural material in the vicinity.

6. Fatehpur V (Baichbal valley), a surface site with multiple localities.

7. and 8 Teggihalli II and Mudnur X (Baichbal valley) two surface sites with small bifaces.

9. Isampur Quarry (Hunsgi valley) an excavated quarry site on a weathered limestone bed (Paddayya *et al.* 1999).

AIMS AND METHODS OF THE PRESENT STUDY

The evidence suggests that over time there may be an increase in the sophistication of Acheulean bifaces. This study aims to assess variation in biface refinement and technology from several sites in one valley, with similar raw material usage patterns. Following previous authors (e.g. White 1998; Milliken 2001), thickness to breadth ratio was used as the indicator of refinement. The thinner a tool is the sharper it is likely to be. The thickness to breadth ratio is also an indicator of the length of the cutting edge as the relatively wider a tool is, the longer its cutting edge. Thickness to breadth ratio reflects the wieldiness of a tool, as the thinner a tool is the lighter it is and the easier it is to grip, making it easier to manoeuvre and apply continuous pressure. Having a secure grip is also particularly important for a butchery tool as it is liable to become slippery as blood, fat and other animal products adhere to the tool during butchery (McNabb 2005). Finally producing a thin tool is possibly the most challenging aspect of biface manufacture: producing a thin biface often requires the production of a large, thin flake blank; it requires sufficiently invasive flakes to be struck to thin the piece whilst maintaining adequate width for an extensive cutting edge (Callahan 1979); and the thinner a biface is, the more liable it is to break due to end-shock during manufacture; therefore thickness to breadth ratio is also an indicator of the skill of the knapper who produced the biface. Thickness and breadth were measured not as the absolute distance between the two points of maximum protrusion but as the distance between these points in only the required dimension as if a box were drawn around the biface.

A number of technological variables were measured in order to ascertain what underpins biface refinement. The overall size of flake scars is an indicator of the amount of force used to remove the flake and therefore the delicateness of the knapping. Producing light flakes with small bulbs of percussion also helps to create a straight edge. The production of a straight edge and symmetrical edge is also aided by small marginal trimming flakes often employed by knappers at the end of bifacial production. In order to ascertain these flake variables the length and width of the two most complete flake scars on each surface of the biface was measured.

A successful flake removal will end in a feather termination, leaving a smooth surface on the biface amenable to further working. However flakes can also terminate more abruptly in hinge or step terminations which create an uneven surface and make future flake removals problematic. Flaking success was calculated as the number of aberrant terminations (hinge and step terminations) divided by the total number of flake scars. The invasiveness of flaking, as mentioned above, is a further indicator of knapping skill, therefore this was measured using Clarkson's (2002) index. This involves dividing each surface of a biface into 8 segments and giving each segment a score depending on how invasive

the retouch is in that segment. A score of 0 means no retouch, a score of 0.5 denotes retouch that does not extend more than halfway from the lateral margin to the mesial point of the tool and a score of 1 indicates retouch that does extend more than halfway to the to the centre. The score is then averaged over the 16 segments of the biface yielding an index of invasiveness. The percentage of cortex remaining on the piece reflects the blank form and the intensity of working, both of which have been argued to affect biface refinement (White, 1995), therefore this variable was estimated to the nearest 5%. Allometric variation has been argued to significantly affect biface variation (Crompton & Gowlett, 1993), therefore the bifaces were weighed to represent their size. The above variables were measured on a total of 350 bifaces. It should be noted that throughout the following analyses the results for Yediyapur IV and Isampur Quarry are preliminary, as data was only collected on a sample of the bifaces from these sites and not the whole assemblage.

ANALYSIS AND RESULTS

In order to characterise the level of refinement in each assemblage the mean values of thickness to breadth ratios are displayed below.

Figure 12.1 shows that bifaces from Mudnur VIII, Isampur, Hunsgi VI and Hunsgi V tend to be less refined than those from Fatehpur V, Yediyapur VI, Yediyapur IV, Teggihalli II and Mudnur X. Regression analyses were conducted on the thickness to breadth ratio to determine which technological factors had a significant relationship with refinement. The following six variables had no significant relationship with refinement: cortex percent, the index of invasiveness and flaking success. The following variables were found to have a significant relationship with refinement: flake scar width to length ratio, flake scar area and weight. The mean values of the variables were plotted by assemblage to see if the site-wise pattern of variation was replicated in these technological variables.

Figure 12.2 shows that the less refined assemblages of Isampur, Mudnur VIII, Hunsgi VI and to a lesser extent Hunsgi V are larger than the more refined assemblages. Yediyapur IV is an anomaly in this case because the sample size is small, only 12 bifaces, and this includes two giant bifaces so the mean is distorted. The two Yediyapur 'giants' are the largest bifaces in the entire study at 2147 g and 287.52 mm long and 4433 g and 321.24 mm, but the rest of the Yediyapur IV assemblage is small. This may be significant as it has been suggested that 'giant' bifaces, being rather unwieldy, may have had a largely symbolic function (Oakley 1981).

Figure 12.3 replicates the pattern of size, flake scar density and refinement whereby the four cruder assemblages are finished with larger flakes, indicating greater percussive force was used in finishing the cruder bifaces.

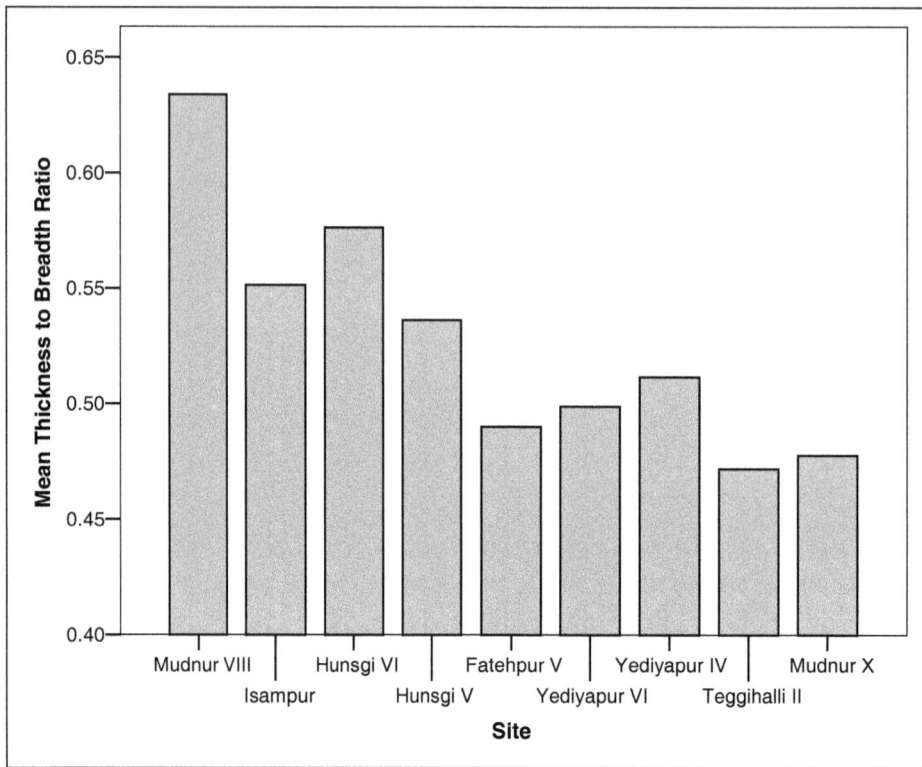

Fig. 12.1. Thickness to breadth ratio by site

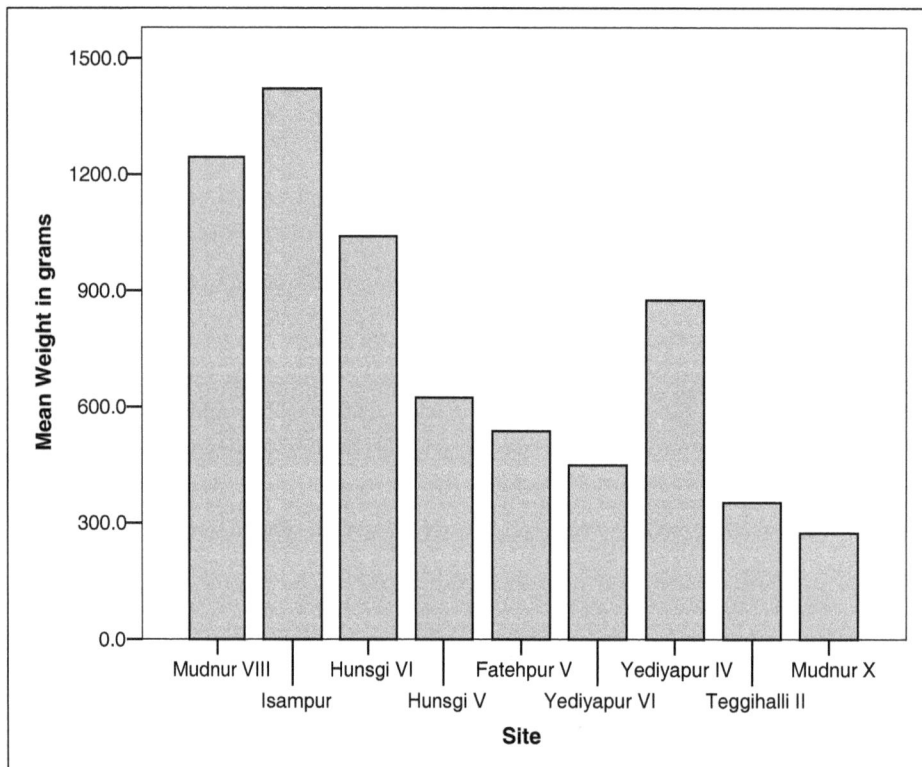

Fig. 12.2. Mean biface weight in grams by site

Figure 12.4 illustrates that the shorter, wider flakes that are produced by soft hammer or delicate working, tend to occur at the sites on the right of the graph with the more refined assemblages, thereby replicating the pattern shown in figure 12.3. The data show clear assemblage wise patterning with bifaces from the more refined assemblages being lighter and having smaller and relatively wider flake scars.

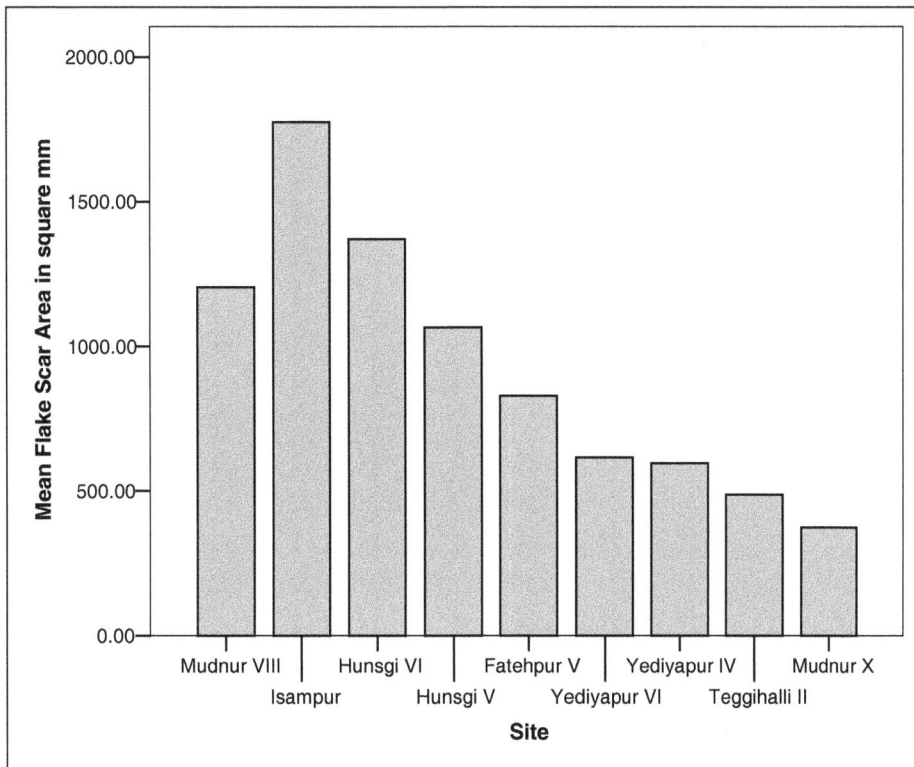

Fig. 12.3. Mean flake scar area by site in mm^2, for the last four flake scars removed from the piece

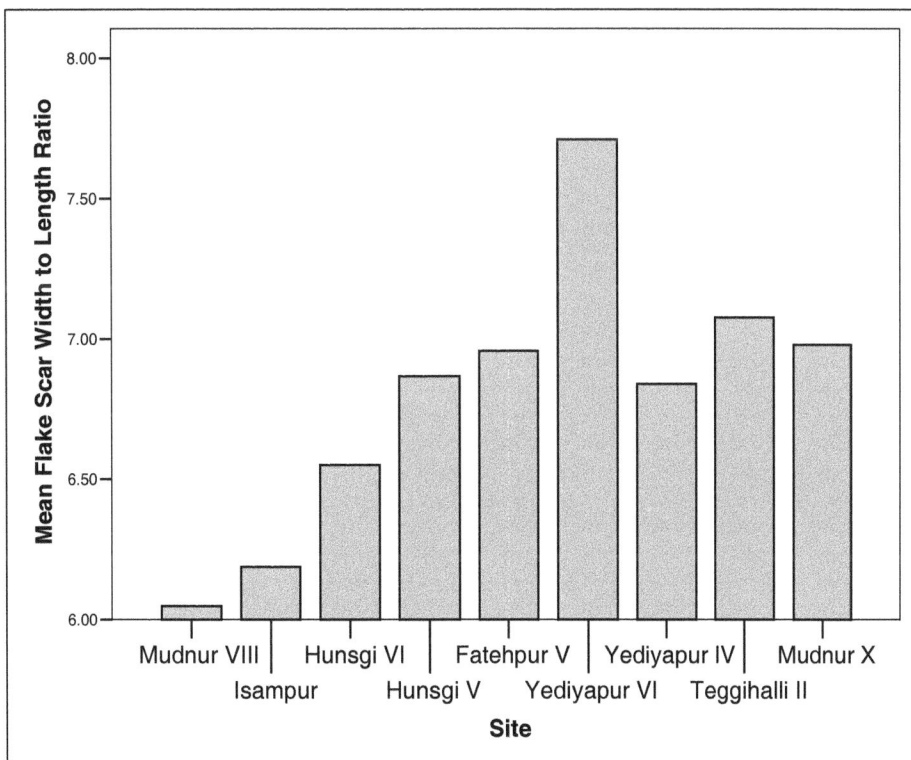

Fig. 12.4. Flake scar width to length ratio by site for the last four flakes removed from the bifaces

DISCUSSION

It has been argued that allometric patterns of bifacial variation, in particular the pattern of long, narrow bifaces versus short, broad ones, are attributable to differing levels of rejuvenation (McPherron 2000). This data supports this hypothesis as figures 12.1 and 12.2 show that assemblages with smaller bifaces also contain relatively wider ones. Figures 12.3 illustrates that assemblages with smaller bifaces tend to have more delicately worked bifaces as might be expected if they were near to the end of the reduction sequence. Figure 12.4 shows that the assemblages with small, refined bifaces are also characterised by the short broad marginal trimming flakes which represent the final stage of bifacial working. This hypothesis is lent further support from the fact that of the two assemblages with largest bifaces, Isampur is a known quarry site, while Mudnur VIII is suspected to be a cache implying that bifaces from both these sites are near to the beginning of the use-lives and would not have undergone many resharpening episodes.

Before we commit to the conclusion that the variation described above is the result of differential rejuvenation, we should consider that resharpening models are based on modern behaviour and may not be applicable to Acheulean hominins. During the 1.25 million years or so, of the Acheulean, hominin brain size undergoes its biggest expansion, increasing from around 850cc in species such as *Homo ergaster,* to 1300cc in species such as *Homo heidelbergensis* (Ruff *et al.* 1997), therefore we should not shy away from recourse to socio-cognitive explanations of Acheulean variation. Given the broad range of dates for the Acheulean in the Hunsgi-Baichbal Valley (1.2 mya for Isampur Quarry to 285 kya for Teggihalli II), socio-cognitive differences between the hominins who manufactured different assemblages are plausible. In this way the small, refined bifaces made with delicate flaking may have been produced by more skilful hominins. At Isampur Quarry hominins tended to ignore the thinnest slabs of limestone (less than 4 cm thick), when procuring handaxe blanks (Petraglia *et al.* 2005; Shipton et al in press), perhaps because their inability to control their percussive force made them prone to end-shock breakage of thinner pieces.

We have offered rejuvenation and socio-cognitive variation as two alternative explanations of the patterns in our data, in reality rejuvenation and socio-cognitive differences may not be independent as differences in planning capacity between different hominins, may underpin the differences in rejuvenation.

Acknowledgements

The authors would like to thank Dr Chris Clarkson for helpful comments and Dr Richa Jhaldyial for help in fieldwork and understanding the geology of the Hunsgi-Baichbal Valley. The Allen, Meek and Read scholarship; the Prehistoric Society Tessa and Mortimer Wheeler fund; the Smithsonian Human Origins Program; the Leakey Foundation; the University Grants Commission, New Delhi; and Deccan College, Pune, are all gratefully thanked for their financial support.

References

BLACKWELL, B.A.B., FEVRIER, S. BLICKSTEIN, J.I.B., PADDAYYA, K., PETRAGLIA, M.D., JHALDIYAL, R., and SKINNER, A.R. (2001) ESR dating of an Acheulian Quarry site at Isampur, India. *Journal of Human Evolution* 40.

BYE, B., BROWN, F.H., CERLING, T.E. and MCDOUGALL, I. (1987) Increased age estimate for the Lower Palaeolithic hominid site at Olorgesailie, Kenya. *Nature* 329: 237-239.

CALLAHAN, E. (1979). The basics of biface knapping in the Eastern Fluted Point Tradition: A manual for flintknappers and lithic analysts. *Archaeology of Eastern North America* 7:1172.

CARBONELL, E., MOSQUERA, M., OLLE, A., RODIRGUEZ, X.P., SALA, R., VERGES, J.M., ARSUAGA, J.L. and BERMUDEZ DE CASTRO, J.M. (2003) Did the earliest mortuary practices take place more than 350.000 years ago at Atapuerca? *L'Anthropologie* 107: 1-14.

CLARK, J.D. (1994) The Acheulian industrial complex in Africa and elsewhere In Corrucini, R.S. and Ciochon, R.L. (eds.) *Integrative pathways to the past* pp 451-470 Englewood Cliffs, NJ: Prentice-Hall.

CLARK, J.D. (2001) Variability in primary and secondary technologies of the Later Acheulian in Africa. In Milliken, S. & Cook, J. (eds.) *A very remote period indeed: papers on the Palaeolithic presented to Derek Roe*. Oxford: Oxbow Books.

CLARKSON, C. (2002) An index of invasiveness for the measurement of unifacial and bifacial retouch: a theoretical, experimental and archaeological verification *Journal of Archaeological Science* 29: 65-75.

CROMPTON, R.H. and GOWLETT, J.A.J. (1993) Allometry and multi-dimensional form in Acheulean bifaces from Kilombe, Kenya. Journal of Human Evolution 25:175-199.

DEBONO, H. and GOREN-INBAR, N. (2001) Note of a link between Acheulian handaxes and the Levallois method. *Journal of the Israel Prehistoric Society* 31: 9-23.

DOMINGUEZ-RODRIGO, M., SERRALONGA, J., JUAN-TRESSERRAS, J., ALCALA, L., LUQUE, L. (2001) Woodworking activities by early humans a plant residue analysis on Acheulian stone tools from Peninj (Tanzania) *Journal of Human Evolution* 40: 289-299.

GAMBLE, C. (1995) The earliest occupation of Europe: the environmental background, in Roebroeks, W. and van Kolksholten, T. (eds.) *The Earliest Occupation of Europe* pp. 279-295, Leiden: European Science Foundation and University of Leiden.

GOREN-INBAR, N. and SARAGUSTI, I. (1996) An Acheulian bifacial assemblage from Gesher Benot Ya'aqov: indications of African affinities. *Journal of Field Archaeology* 23: 15-30.

HALSTEAD, L.B. (1982) *Hunting the Past: Fossils, Rocks, Tracks and Trails; the Search for the Origin of Life* Book Club Associates: London.

ISAAC, G.Ll. (1977) *Olorgesailie: the archaeology of a Middle Pleistocene lake basin in Kenya.* Chicago: Chicago University Press.

JONES, P.R. (1980) Experimental butchery with modern stone tools and its relevance for Palaeolithic archaeology, World Archaeology 12: 153-175.

KEELEY, L.H. (1980) *Experimental Determination of Stone Tool Uses: a Microwear Analysis* Chicago: University of Chicago Press.

MCNABB, J. (2005) Hominins and the Early-Middle Pleistocene transition: evolution, culture and climate change in Africa and Europe. In Head, M.J. and Gibbard, P.L. (eds) *Early-Middle Pleistocene Transitions: The land-ocean evidence.* Geological Society, London, Special Publications 247: 287-304.

MCPHERRON, S.P. (2000) Handaxes as a measure of the mental capabilities of early hominids. *Journal of Archaeological Science* 27:655-663.

MILLIKEN, S. (2001) Acheulian handaxe variability in Middle Pleistocene Italy: a case study. In Milliken, S. and Cook, J. (eds.) *A very remote period indeed: papers on the Palaeolithic presented to Derek Roe* pp. 160-173. Oxford: Oxbow Books.

MITCHELL, J.C. (1997) Quantitative image analysis of lithic microwear on flint handaxes. *USA Microscopy and Analysis* 26:15-17

OAKLEY, K.P. (1981). Emergence of higher thought 3.0-0.2 Ma B.P. *Phil. Trans. R. Soc. London B* 292:205-211.

PADDAYYA, K. (1977) An Acheulian occupation site at Hunsgi, peninsular India: a summary of the results of two seasons of excavation (1975-76) *World Archaeology* 8: 344-355.

PADDAYYA, K. (1979) Excavation of a new Acheulian occupation site at Hunsgi, peninsular India. *Quartar* 29-30: 139-155.

PADDAYYA, K. (1982) *The Acheulian culture of the Hunsgi Valley (Peninsular India): a settlement system perspective.* Poona: Deccan College Postgraduate and Research Institute.

PADDAYYA, K. (1987) Excavation of a new Acheulean occupation site at Yediyapur, Peninsula India. *Anthropos* 82: 610-614.

PADDAYYA, K. (2001) The Acheulian culture project of the Hunsgi and Baichbal valleys, peninsula India. In *Human Roots: Africa and Asia in the Middle Pleistocene*, L. Barham and Kate Robson-Brown (eds.) pp. 235-258, Bristol: Western Academic Press.

PADDAYYA, K. and PETRAGLIA, M. (1995) Natural and Cultural Formation Processes of the Acheulian sites of the Hunsgi-Baichbal Valleys, Karnataka. In Wadia, S., Korrisettar, R. & Kale, V.S. (eds.) *Memoirs of the Geological Society of India 3, Quaternary Environments and Geoarchaeology of India: Essays in Honour of Professor S.N. Rajaguru* pp.333-351. Geological Society of India: Bangalore.

PADDAYYA K., BLACKWELL B.A.B., JHALDIYAL R., PETRAGLIA M.D., FEVRIER S., CHADERTON II D.A., BLICKSTEIN J.I.B. and SKINNER A.R. (2002) Recent findings on the Acheulian of the Hunsgi and Baichbal valleys, Karnataka, with special reference to the Isampur excavation and its dating *Current Science 83:5.*

PADDAYYA, K., JHALDIYAL, R., PETRAGLIA and M.D. (1999) Geoarchaeology of the Acheulian Workshop at Isampur, Hunsgi Valley, Karnataka. *Man and Environment* 24:95-100.

PADDAYYA, K JHALDIYAL, R and PETRAGLIA, M.D. (1999) Geoarchaeology of the Acheulean workshop at Isampur, Hunsgi valley, Karnataka *Man and Environment* 24(1), 167-184.

PETRAGLIA, M.D., SHIPTON, C. and PADDAYYA, K. (2005) Life and Mind in the Acheulean: a case study from India. In *The Hominid Individual in Context: Archaeological investigations of Lower and Middle Palaeolithic landscapes, locales and artefacts.* Gamble, C. & Porr, M. (eds.) pp. 197-219. Routledge: London.

PITTS, M. and ROBERTS, M. (1998) *Fairweather Eden: Life in Britain Half a Million Years Ago as Revealed by the Excavations at Boxgrove* Random House: London.

RAYNAL J.P., L. MAGOGA, F. SBIHI-ALAOUI and D. GERAADS, (1995). The Earliest Occupation of Atlantic Morocco: the Casablanca Evidence pp. 255-262. In *The Earliest Occupation of Europe.* (ed. by W. Roebroeks & T. Van Kolfschoten) Leiden: University of Leiden.

RAYNAL J.P. and TEXIER J.P (1989). Découverte d'Acheuléen Ancien Dans la Carrière Thomas 1 à Casablanca et Problème de L'ancienneté de la Présence Humaine au Maroc. C.R.A.S. Paris 308, 1743-1749.

ROBERTS, M.B. and S.A. PARFITT. (1999) *Boxgrove: a Middle Pleistocene hominid site at Eartham Quarry, Boxgrove, West Sussex.* London: English Heritage.

ROCHE, H. and KIBUNJIA, M. (1994) Les sites archeologiques plio-pleitcoenes de la Formation de Nachukui, West Turkana, Kenya. *Comptes Rendus de l'Acadmie des Sciences de Paris* 318, serie II: 1145-1151.

ROE, D.A. (1971) The Chelles-Acheul culture of East Africa. In Leakey, M.D. (ed.) *Olduvai Gorge: Excavations in Beds I and II, 1960-1963.* Cambridge: Cambridge University Press.

ROE, D.A. (2001) The Kalambo Falls large cutting tools: a comparative metrical and statistical analysis. In *Kalambo Falls prehistoric site. Vol. 3, The earlier cultures: Middle and Earlier Stone Age*, J. Desmond Clark and Julie Cormack (eds.). Cambridge: Cambridge University Press.

ROEBROEKS, W. (1994) Updating the earliest occupation of Europe *Current Anthropology* 35: 301-305.

RUFF, C.B., TRINKHAUS, E. & HOLLIDAY, T. (1997) Body mass and encephalisation in Pleistocene *Homo. Nature*, 387: 173-176.

SARAGUSTI, I., KARASIK, A., SHARON, I., SMILANSKY, U. (2005) Quantitative analysis of shape attributes based on contours and section profiles in artefact analysis. *Journal of Archaeological Science* 32: 841-853.

SARAGUSTI, I., I. SHARON, O. KATZENELSON and D. AVNIR (1998). Quantitative Analysis of the Symmetry of Artefacts: Lower Paleolithic Handaxes. *Journal of Archaeological Science* 25:817-825.

SHIPTON, C., PETRAGLIA, M. and PADDAYYA, K. (in press) Inferring aspects of Acheulean Sociality and Cognition from Biface Technology. In Brooke, B. & Adams, B. (eds.) *Lithic Materials and Palaeolithic Societies.* Blackwell.

SZABO, B.J., MCKINNEY, C., DALBEY, T.S. and PADDAYYA, K. (1990) On the Age of the Acheulian Culture of the Hunsgi-Baichbal Valleys, Peninsular India. *Bulletin of the Deccan College Post-Graduate and Research Institute* 50: 317-321.

WHITE, M.J. (1995) Raw materials and biface variability in southern Britain: a preliminary examination *Lithics* 15: 1-20.

WHITE, M.J. (1998) On the significance of Acheulean biface variability in southern Britain. *Proceedings of the Prehistoric Society* 64: 15-44.

WHITE, M.J. and ASHTON, N.M. (2003) Lower Palaeolithic core technology and the origins of the Levallois method in NW Europe *Current Anthropology* 44: 598-609.

WYMER, J. (1982) *The Palaeolithic Age* New York: St. Martin's.

WYNN, T. (2002) Archaeology and cognitive evolution *Behavioural and Brain Sciences* 25:4.

EVOLUTIONARY TRANSITIONS AND CO-EVOLUTIONARY DYNAMICS IN BIOLOGY AND IN CULTURE

Mónica TAMARIZ

Language Evolution and Computation Research Unit, Department of Linguistics and English Language, The University of Edinburgh, monica@ling.ed.ac.uk

Abstract: This paper presents a Darwinian framework to study culture that formalises interactions between public and private, ontogenetic and phylogenetic as well as individual and social aspects of cultural evolution and transmission. It also compares and contrasts evolutionary milestones in the emergence of culture with major transitions in the evolution of life. We define two related processes in the evolution of culture: the cumulative encoding of innovative information into public culture and the ontogenetic development of cultural competences that allow humans to access and use that information. We claim that the capacity to create, learn and use symbols is a key factor underlying those processes.
Keywords: Cultural evolution; evolutionary transition; public culture; private culture

Résumé: On present un cadre Danwinien pour etudier la culture qui formalise les interactions entre des aspects de l'evolution et la transmission culturelle public et prives, ontogenetiques et phylogenetiques, individuels et sociaux. Ce cadre theoretique compares les transitions evolutives dans l'emergence de la culture avec cuus dans l'evolution biologique. On definira deux proces: la codification cumulative d'information nouvelle dans la culture publique et le developpement ontogenetique des compeetences culturelles qui nous permettent d'acceder et de se servir de cette information. Ces proces culturels sont possibles grace a la capacite de creer, aprendere et user des symbols.
Mots clés: evolution culturelle; transition evoultive; culture publique; culture privee

The aim of this paper is to present an evolutionary framework for culture informed by the elements and mechanisms of selection dynamics in biology. Natural selection was initially and mainly formulated to explain the evolution of biological species (Darwin 1859), and its application to cultural phenomena was realised by Darwin himself (Darwin 1871). Evolutionary frameworks for culture have been proposed in several fields (Dawkins 1976; Cavalli-Sforza and Feldman 1981; Boyd and Richerson 1985; Dennett 1995; Mace et al. 2005; Croft 2000; Mufwene 2001; Brighton, Smith and Kirby 2005; Schumpeter 1934; Nelson and Winter 1982). The remainder of this section offers an overview of evolution. Then we focus on biology, particularly on the evolutionary transition that gave rise to Darwinian dynamics and the phylogenetic tree of life. The main section of the paper analyzes a similar transition that led to co-evolutionary dynamics within culture. Specifically, we describe two levels of cultural selection, one related to function, where meanings evolve, and another related to form, where the public culture evolves. The final sections highlight the differences and similarities between evolution in biology and in culture by focusing on information flows and on causality in selection dynamics and present some implications of our co-evolutionary framework for the study of culture.

Evolution is a process whereby inheritable features arise and spread in populations. It may happen by random, undirected drift, but the focus of this paper is evolution by selection. Hull (2001) defines selection as "repeated cycles of replication, variation and environmental interaction so structured that environmental interaction causes replication to be differential". In selection systems, "like begets like", but interaction with the environment yields "descent with modification" so that over the generations, the information in a population reflects the structure of its environment. The following paragraphs define and illustrate the elements (the unit of replication and the vehicle) and processes (replication, variation and adaptation) of selection. Fig.13.1. illustrates a selection system.

Hull (1988) defined the unit of replication as "the entity that passes on its structure largely intact in successive generations". Dawkins (1976) emphasized the role of genes as entities that contain the information that is passed on during replication, but we follow Williams' (1992) view that the unit of replication is defined by its information content. Information can be defined for our purposes as any pattern that influences the formation or the transformation of other patterns. Several units of replication have been proposed in culture: in the economy, routines and rules (Nelson and Winter 1982); in the evolution of science, beliefs, goals and methodologies (Hull 1988); in linguistics, structural linguistic features (Croft 2000, Mufwene 2001, Tamariz 2006).

A second element of Hull's (1988) general account of selection is "the entity that interacts as a cohesive whole with its environment". Dawkins (1976) called this the "vehicle". In biology, the vehicle is the phenotype that develops when the genetic information unfolds in an environment. In the economy, Knudsen and Hodgson (2004) propose that firms are vehicles for the replication of habits and routines. In Hull's (1988) evolution of scientific knowledge, the vehicles are the scientists. In some linguistic models, vehicles are the speaker and his grammar (Croft 2000, Mufwene 2001, Tamariz 2006).

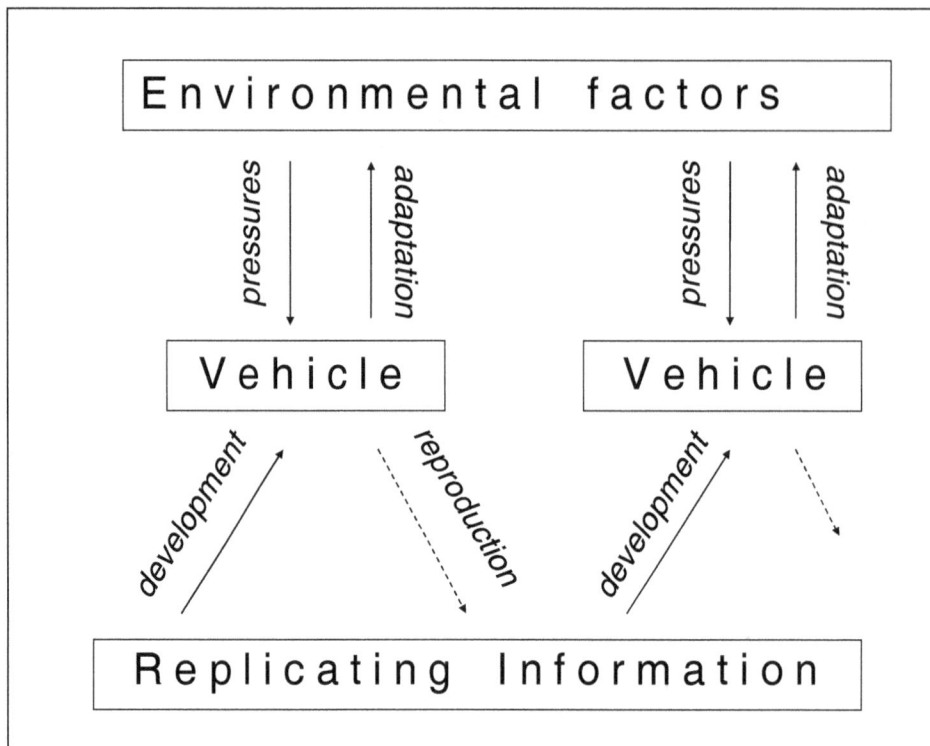

Fig. 13.1. The elements of a Darwinian selection system

Replication is related to copying information. Three criteria define replication in an evolutionary system: causation, similarity and information transfer between original and copy (Hodgson and Knudsen *in press*).

Selection can only occur if there exist inheritable variants of the units of replication with different fitness values. Variation may originate in mutation (a change to the replicating information) and in recombination.

Any non-replicating information that contributes to the self-sustainment of an evolutionary system is considered to be environmental information. Metaphorically speaking, the environmental information constitutes the fitness landscape where replicating information evolves. In that way the environment constrains the possible evolutionary histories of a system, resulting in adaptation. Selection is the process whereby the path of a complex system's evolution is carved in that landscape.

MOLECULAR EVOLUTION

Cyclical chemical reactions are at the heart of a widely accepted hypothesis to explain the origin of life, namely autonomous replication, or "continued growth and division which is reliant on input of small molecules and energy only" (Szostak *et al.* 2001). Primeval life systems present a unit of replication (information about the structure of self-replicating units) and a vehicle (the molecules themselves), and encompass variation, replication and environmental interaction. The inheritance of features

of the molecules is both caused by the replicating information and by the stability of the environment (first the primeval soup and later the intracellular environment, whose homeostasis is tightly regulated and constitutes the first "niche" that self-replicating molecules constructed around themselves and transmitted non-genetically over the lineages). Replication occurs because there is causation, similarity and information transfer when the system keeps producing more of the same networks of molecules. Variation is brought about by random changes in the structure of the component molecules and by recombination of existing ones through horizontal transfer of information. Adaptation results from the inherent interaction between the molecules and their environment.

Apart from horizontal information transfer, early life systems diverge from organismal selection in that they do not include translation mechanisms and therefore the information they contain is only about their own structure (strongly constrained by the adaptive pressure posed by the function of self-replication). In other words, the vehicle is also the repository of the information and consequently, the processes of replication and ontogeny (development) are one and the same.

ORGANISMAL EVOLUTION

Woese (1998) identifies the origin of cells able to translate nucleic acids (DNA, RNA) into proteins as the most important single event in evolutionary history and as one of the great transitions in evolution that are characterized by the appearance of new ways of transmitting information (Maynard Smith and Szathmáry 1995) or by new mechanisms of symbolic representation (Woese 2002). The transition at hand, termed the Darwinian Threshold by Woese, occurred when horizontal transfer of replicating material led to complex cells where the mechanisms of translation evolved. If genes are defined as the stretch of DNA that code for the amino-acid sequence of a protein, then the onset of translation from nucleotide sequences to proteins effectively brought about a new unit of information: the gene. Woese (1998) points out that, with translation in place, vertical transfer of genetic information leads to an increasingly permanent organismal genealogical trace. Speciation in the Darwinian sense begins and genetic information is now amenable to representation by a tree topology.

Replication of genes is mediated by the phenotypes. Mechanisms for variation include mutation, and recombination, which may occur by sexual reproduction or by horizontal gene transfer (the latter is illustrated e.g. by Woese 2000's report of cases of acquired antibiotic resistance in bacteria). Adaptation is observed at the level of the phenotype, with respect to its development and its ability to reproduce in an environment. The informational systematicity between genes and the environment is achieved through natural selection.

THE ORIGIN AND EVOLUTION OF CULTURE

Let us start with a definition by Mesoudi *et al.* (2004): "[Culture is] acquired information, such as knowledge, beliefs, and values, that is inherited through social learning, and expressed in behavior and artifacts". This definition points at two distinct entities: private or cognitive information residing in individual minds and public manifestations of culture expressing the private information. Mesoudi *et al.'s* definition seems to imply that culture comprises private aspects only while public culture is the expression of private culture. Similarly, for Boyd and Richerson (1985), and the memetics literature (Dawkins 1976, Dennett 1995), culture is information stored in human brains. The focus of cultural and social anthropology, on the other hand, is the material or public aspect of culture.

Back to Mesoudi *et al.* (2004)'s definition, culture is not innately or genetically specified, but socially learned during an individual's lifetime. Additionally, public culture is symbolic because the information is "about" something other than the repository of the information, and internally structured because informational entropy is not maximal, and therefore some redundancy (complexity) can be measured. Some animal communication systems share some of these characteristics: birdsong is socially learned and is internally structured but not symbolic (it does not have meaning); apes can use, learn and even categorize symbols (Savage-Rumbaugh *et al.*

1980), but cannot cope with complex syntactic structure; bee dance is symbolic and structured, but not socially learned. The only system that is mostly socially learned, symbolic and structured is human culture, the main subject of this paper. Ultimately we must not forget that culture is limited by the fact that it must, overall, increase human fitness.

CULTURAL TRANSITIONS

We distinguish three states of a population with respect to culture: in the first, there is no communication between individuals; in the second, an unstructured communication system allows transmission of meanings between individuals, and in the third, the communication system is structured. The "Cultural threshold" is positioned between the second and third.

We assume that before culture emerged, our ancestors could entertain thoughts, and that this is also the case with other primates (Hurford 2007). The repository for the replicating information in meanings or thoughts are patterns of neural activity akin to Aunger's (2002) "electric memes". These meanings were locked inside individual brains and may have been produced repeatedly, or replicated, when prompted by external or internal events. We propose that this stage is formally analogous to Woese's pre-Darwinian era in biology: some information is maintained over time (within an individual's lifetime) thanks to stable environmental input (objects and events in the world). Moreover, horizontal transfer of information between meanings within one brain may have been possible through of metaphor and analogy that would have been resulted, for instance, from the increased cognitive fluidity proposed by Mithen (1996).

A first transition occurs when meanings are encoded in a repository different from the neural substrate. The advent of communication is enabled by the evolution of a translation mechanism that allows encoding and decoding between private meanings and public forms, namely symbolic association. (Note that innate symbolic association is present in animal communication systems, but human communication is socially learned). Communication systems bring about two novelties: the replication of meanings between brains and the production of symbols, (public behaviours and artefacts that express meanings), both of which play a crucial role in the emergence of culture.

The transition to culture is marked by the advent of a new cognitive capacity to create and learn new symbolic associations between patterns observed in public culture (which is symbolic itself) and existing or new meanings. This capacity generates a process of structuring, complexification or organization (i.e. evolution) of public forms over time. We can define a second translation process that induces the emergence of a new kind of

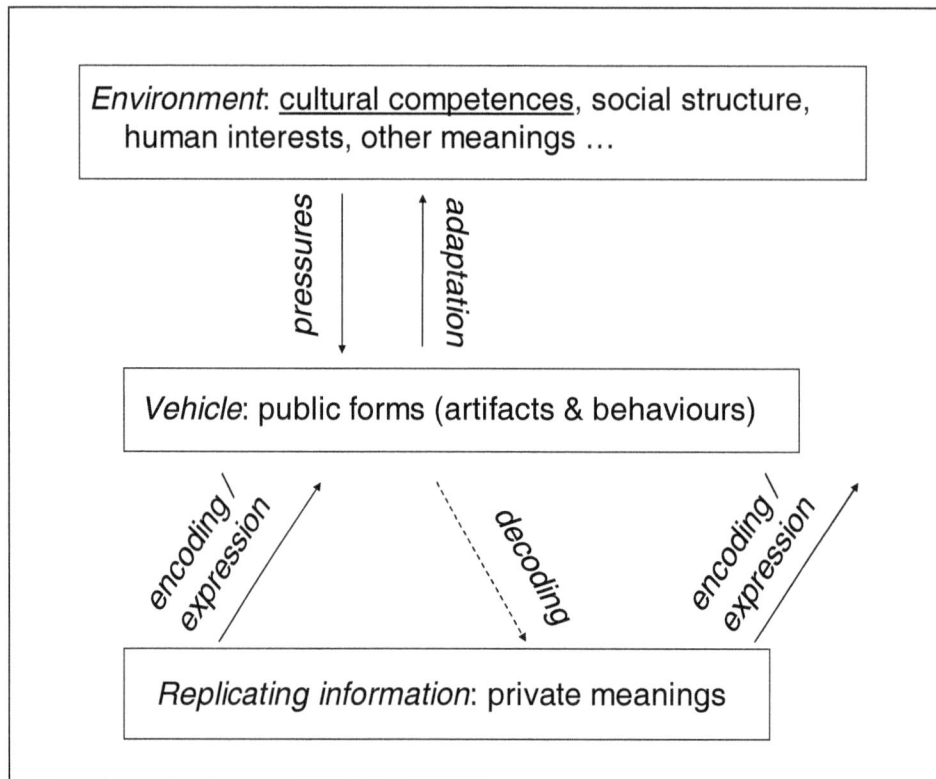

Fig. 13.2. Elements and mechanisms of selection of public forms

information. In a communication system, the sturcutre of public forms is in principle irrelevant (as long as they can be perceived as distinct from each other). We claim that, in a cultural system, an innate, human-specific ability to detect patterns in the structure of public forms and, crucially, to convert those patterns into symbols by assigning them meanings (non-innately) makes the structure of public forms relevant. The new ability results in the ontogenetic development in the brain of each member of a society of a collection of symbolic associations between meanings and forms. We call these mental entities cultural competences, because they allow individuals to encode and decode, to access and to use the information in public culture. These novelties dramatically transform the communication system by making it cumulative. Meanings transcend the spatio-temporal limits of individual brains and lifetimes by finding a new repository in public forms. Public forms become ever more complex because information about their own structure evolves in populations over the generations.

THE EVOLUTION OF PUBLIC FORMS: COMMUNICATION

We now describe a co-evolutionary framework encompassing public and private aspects of culture that includes two selection systems, each with its units of replication, vehicles, environmental interactions, and mechanisms of replication and variation. In a way similar to the above description of molecular and organismal selection, we

will describe selection of public forms (Fig. 13.2) and of competences (Fig. 13.3).

Units of Replication and Vehicles. During communication, public forms are the vehicles that express (that are symbolically associated with) private meanings. The units of replication, meanings, are neurally encoded and, therefore, private.

Replication. Replication happens during communication when a copy or a person's private meaning is produced in another person's brain.

Variation. Variation of meanings can originate in horizontal transfer, or recombination, of information among the meanings residing in the same brain during metaphorical and analogical activity.

Adaptation. Several environmental factors configure the fitness landscape where meanings evolve. First, cultural competences, the conventionalized mappings that allow encoding and decoding between meanings on the one hand and cultural behaviours and artefacts on the other hand, which are explained in detail in the next section. Second, innate biases related to human fitness determine the extent to which different meanings are expressed and attended to by making individuals devote preferential attention, time and resources to certain aspects of the environment, such as the social structure, mating or food. Third, the social structure bears on the fitness of meaning, as it affects the opportunities for communication between

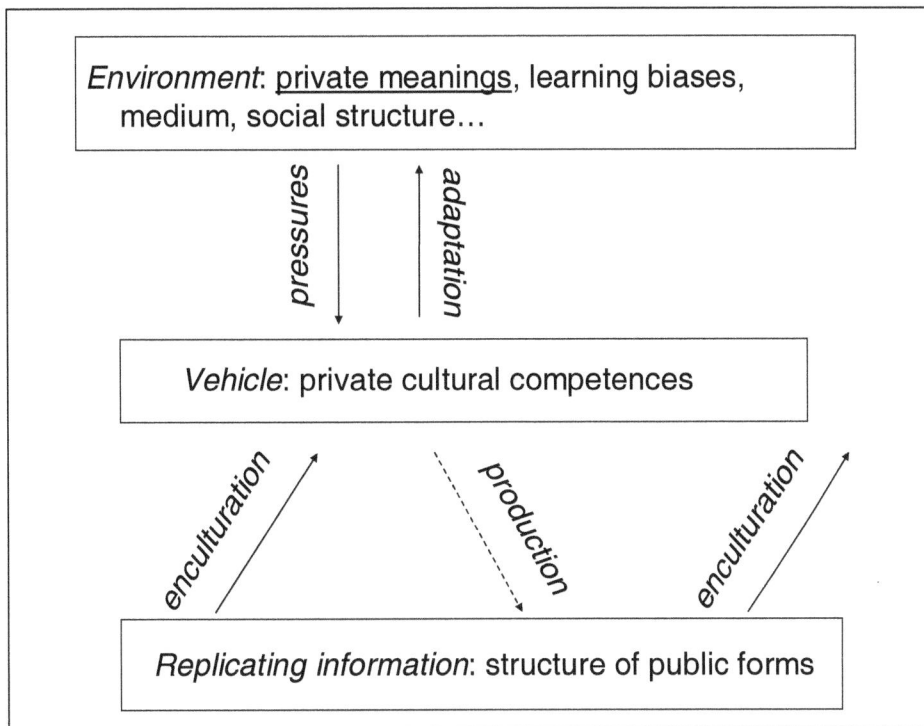

Fig. 13.3. Elements and mechanisms of selection of cultural competences

people. Fourth, other meanings in recipient brains interact with existing meanings, affecting their fitness.

THE EVOLUTION OF COMPETENCES: CULTURE

As we saw above, the transition to organismal evolution in biology took place when translation between nucleic acid strings and proteins became available and brought about a new kind of symbolic information (genetic information). In culture, translation between private meaning and public form brought about the possibility of communication. The cultural threshold (second transitions) was crossed when a new encoding of information emerged. When humans (genetically) evolved the capacity to learn symbolic associations between private meanings and patterns in public culture, competences, or sets of associations between inferred meanings and observed patterns in public culture, began to develop in our brains thanks to exposure to and use of culture. Fig. 13.3 illustrates the resulting selection system.

Units of Replication and Vehicles. Our account of the emergence of culture entails the appearance of a new kind of replicating information: the regularities detected in the structure of public artefacts and behaviours. This new information is symbolically associated to aspects of the meanings that the public forms convey, for instance, the occurrence of "–ed" in English speech is inferred to be associated with past tense. The vehicles for the replication of the new information are (private) human cultural linguistic, social, technological or economic competences, the codes that develop during enculturation, for instance,

through disambiguation across multiple contexts (Smith 2005).

Replication. The replication of structural features of public forms occurs when an individual produces cultural output that has the same structure as the cultural input which contributed to his or her enculturation. For instance, when a person speaks, their output speech has the same structure (phonology, syntax) as the language that elicited the development of her linguistic competence.

Variation. Variation in the pool of public culture features may originate in imperfect replication (mutation) and in recombination (e.g. combination of form feature information from different sources during enculturation).

Adaptation. The environmental factors that determine the fitness of the replicating information about the structure of public forms include the social structure and innate learning biases such as the human capacity to learn from repeated exposures (Smith 2005), the ability to participate in social interaction involving shared attention (Tomasello 2003) and the capacity to create and manipulate new symbolic associations between patterns inferred from public culture and new meanings, that is, to turn public culture information into symbols (Deacon 1997).

CO-EVOLUTIONARY INTERACTIONS IN CULTURE

We have proposed that there are two evolutionary systems in culture. The first is concerned with the transmission of private meanings between brains by means of cultural forms during individual acts of communication. The second is concerned with the transmission of the structure of public forms over the generations by means of competences. The two are intimately connected: the development of cultural competences (enculturation) is a prolonged process involving repeated single interactions in which meanings are encoded into and decoded from public forms. We now describe three interactions that characterise their co-evolution. First, competences are needed for the replication of meanings, i.e. for encoding and decoding between private meanings and public cultural forms. Second, private meanings are needed for the development of competences: This development involves establishing symbolic mappings between structural features of forms and aspects of meanings. Third, conversely, the extraction of structural patterns from public culture may result in the creation of new meanings, as patterns that are noticed may be associated with a consistently co-occurring meaning. Fourth, public forms as vehicles for meanings are the repositories for the replicating information of the system of competences (structural features of public forms).

EVOLUTION OF LIFE, EVOLUTION OF CULTURE

We have now a detailed account of two selection systems in life and in culture and are in a position to compare and see interactions between both. The most relevant similarity is the fact that both in culture and in biology we find two co-evolving systems, the second of which results from an evolutionary transition characterized by a mechanisms able to extract information from a structure that so far functioned as a vehicle for other information patterns.

The co-evolutionary relationships between genes and culture have been the subject of extensive study (e.g. Durham 1991, Cavalli-Sforza and Feldman 1981, Boyd and Richerson 1985, Lumdsen and Wilson 1981). During the evolution of our cultural niche (Odling-Smee, Laland and Feldman 2003) humans have become increasingly reliant on culture, which has deeply transformed the fitness landscape of human genes. The information (knowledge, meanings) that humans use is increasingly encoded in public cultural repositories, and correspondingly less in private neural repositories. Natural selection's adaptation to this shift is a neural environment capable of translating between private meanings and public forms. Public culture has thus relieved the human brain of the pressure to store large amounts of information. With less pressure to encode meanings privately, neural resources can be devoted to competences, the interface between currently relevant meanings and public culture. For example, the invention of writing relieved storytellers from the need to carry the stories in their heads, but required them to learn to read. Cultural evolution has entailed a process of downloading

information from brains onto public culture, while brains have adapted to house interfaces that help us access and use parts of the information encoded in culture efficiently and only as required. For a new competence to be evolutionarily stable, the amount of information it makes available (by accessing it from public culture) must be greater than the amount of "memory space" it takes in the brain, or the information it could hold in the same amount of neural resources.

One important difference between culture and biology concerns reproduction. In culture, the structural public culture information that recombines in each new individual competence does not come from two parents, as in biological sexual reproduction, but from a multitude of other individuals. Indeed, we can learn from older, younger and contemporary individuals. This might preclude the existence of traceable lineages of public cultural information; however, because of the asymmetrical learning between human generations (i.e. children tend to learn from parents more than vice versa), information transmission in the system of competences such lineages become traceable. This is related to another departure of the form selection system from the paradigmatic case of organismal natural selection. In sexual organisms, fertilization is the process whereby genetic material (information) from an egg and a sperm fuse to form a new genotype. In cultural competences, gathering replicating information is an extended process that continues throughout an individual's lifetime. Learning during enculturation is incremental, which means that the patterns learnt by one individual become more robust as he is exposed to more exemplars. Early exposures have a greater impact on the development of competences and later exposures have decreasing impact, contributing to a flow of information down the human generations where younger individuals are net recipients and older individuals are net contributors. This unidirectional net flow of information results in a mostly vertical transmission that underlies a stable genealogy of the information about the structure of public cultural forms.

PREDICTIONS AND EXTENSIONS OF THE CO-EVOLUTIONARY FRAMEWORK

The co-evolutionary dynamics for culture we have described produces predictions that could be tested empirically or with computer simulations concerning the origin, workings and evolution of cultural systems.

The complexity of an evolving system increases over time through an accumulation of frozen accidents (Gell-Mann 1994). One prediction stemming from this fact is that increased complexity in the two proposed co-evolving systems should boost each other's complexity in three ways: first, unintended information may be extracted from the cultural environment, leading to complexification over the generations of cultural competences; second, more complex competences may encode and decode more complex meanings into forms; third, more complex forms may contain more complex structural information and fourth, more complex meanings pose pressure for more complex vehicles to encode them. This hypothesis has been successfully tested with a computer model (Tamariz and Vogt, in preparation); further evidence could be gathered from examining the rate and timing of change of private and public culture in various domains, such as archaeology, the economy, technology, anthropology or linguistics. Moreover, the synergistic complexification of culture may have posed a pressure for the complexi-fication of the neural substrate.

The co-evolutionary framework can be applied at other levels of analysis within culture, which can be illustrated with an example from information technology: the evolution of information contained in computer files and in the software and hardware used to transmit that information. Evolutionary transitions in culture happen when new repositories of information become available and new competences evolve to access and use the information in the new repositories. The equivalent evolutionary transitions in IT happen when new ways of storing information are used (e.g. files used to be stored in individual computers, now they can be stored on the Internet, disks etc.) and new ways to access and use that information emerge (e.g. increased processing power to compress and decompress files, increased connectivity and bandwidth to upload and download them from the new repository). In each such transition the pressure on earlier storage devices is eased: we saw earlier how the advent of the printed word relieved individuals from the pressure to commit information to memory; similarly, the Internet may store vast amounts of information that would not fit into a single computer. The process that co-evolves with this is an ever-increasing complexification of the cognitive competences and the technologies that allow people and computers, respectively, to encode informa-tion onto public domains and download it as required, which is, effectively, what humans do in our lives as cultural beings (see Clark 2003). This trend, in turn, poses a pressure towards the complexification of the substrate for those competences, be they neural or computer hardware, which is attested by the evolution of the human brain and of the communication apparatus and of information technology.

CONCLUSION

A parallel account of the evolutionary dynamics of culture and biology has revealed that despite obvious differences, fundamental similarities can be observed between the two. Furthermore, these general principles may be applicable to other domains. These commonalities may be characteristic of adaptive complex systems undergoing transitions prompted by new translation mechanisms. We have made the following specific claims about cultural evolution:

The evolutionary dynamics of the systems of cultural forms and competences are analogous in some fundamental ways to molecular and organismal evolution in biology. In both cases, a transition occurs when a new kind of symbolic information previously present is processed by a translation system that ultimately leads to replication of the new information.

Culture comprises two kinds of information: neurally-encoded private meanings and information about the structure of public cultural forms. They define two selection systems that evolve at different rates through different mechanisms but nevertheless are integrated within one co-evolutionary unit, as they provide fundamental evolutionary elements and mechanisms for one another. This account of culture can be inscribed within a wider gene-culture co-evolutionary framework. Additionally, co-evolutionary dynamics can be applied to complex interactions at other levels within culture.

The result of cultural co-evolution is that private meanings can be encoded in the virtually unlimited distributed repository that is public culture. Individual cultural competences are the interfaces that allow individuals to interact with public culture environment as and when needed. As new repositories of information emerge, complexifying public culture, the competences that process that information also become increasingly complex. This in turn poses a pressure on natural selection of genetic information for the complexification of neural resources.

Acknowledgements

Writing of this paper has benefited from the financial support of ESRC Postdoctoral Fellowship nr. R39681.

References

AUNGER, R. (2002) The electric meme. New York: Free Press.

BOYD, R., RICHERSON P.J. (1985) Culture and the evolutionary process. Chicago: University of Chicago Press.

BRIGHTON, H., SMITH K., KIRBY S. (2005) Language as an evolutionary system. Physics of Life Reviews 2(3): 177–226.

CAVALLI-SFORZA, L.L., FELDMAN M.W. (1981) Cultural transmission and evolution: A quantitative approach. Princeton: Princeton University Press.

CLARK, A. (2003) Natural-born cyborgs: Minds, techno-logies and the future of human intelligence. New York, Oxford University Press.

CROFT, W. (2000) Explaining language change: an evolutionary approach. Harlow: Longman.

DARWIN, C. (1859) The origin of species by means of natural selection. London: John Murray.

DARWIN, C. (1871) The descent of man and selection in relation to sex. London: John Murray.

DAWKINS, R. (1976) The selfish gene. Oxford: Oxford University Press.

DEACON, T. (1997) The symbolic species: the co-evolution of language and the human brain. London: Penguin Books.

DENNETT, D.C. (1995) Darwin's dangerous idea. New York: Simon & Schuster.

DURHAM, W.H. (1991) Coevolution: Genes, culture and human diversity. Stanford CA: Stanford University Press.

GELL-MANN, M. (1994) The Quark and the Jaguar. New York: Freeman.

HODGSON, G.M., KNUDSEN T. (in press) Dismantling Lamarckism. Journal of Evolutionary Economics.

HULL, D.L. (1988) Science as process: An evolutionary account of the social and conceptual development of science. Chicago: University of Chicago Press.

HURFORD, J.R. (2007) The origins of meaning (Vol. 1 of Langue in the light of evolution). Oxford: Oxford Univeristy Press.

KIRBY, S. (1999) Function, selection, and innateness. Oxford: Oxford Univeristy Press.

KNUDSEN, T., HODGSON G.M. (2004) The Firm as an interactor: Firms as vehicles for habits and routines. Journal of Evolutionary Economics 14(3): 281–307.

LUMSDEN, C.J., WILSON E.O. (1981) Genes, mind and culture: The coevolutionary process. Cambridge MA: Harvard University Press.

MACE, R., HOLDEN, C.J., SHENNAN S. (2005) The evolution of cultural diversity: A phylogenetic approach. London: UCL Press.

MAYNARD SMITH, J., SZATHMÁRY E. (1995). The major transitions in evolution. Oxford: Freeman.

MESOUDI, A., WHITEN A., LALAND, K.N. (2004) Is human cultural evolution Darwinian? Evidence reviewed from the perspective of 'The Origin of Species'. Evolution 58(1): 1–11.

MITHEN, S. (1996) The prehistory of the mind. London: Thames & Hudson.

MUFWENE, S.S. (2001) The ecology of language evolution. Cambridge MA: Cambridge University Press.

NELSON, R.R., WINTER S.G. (1982). An evolutionary theory of economic change. Cambridge MA: Harvard University Press.

ODLING-SMEE, FJ, LALAND KN, FELDMAN MW (2003) Niche construction: The neglected process in evolution. Princeton: Princeton University Press.

SAVAGE-RUMBAUGH, E.S., RUMBAUGH D.M., SMITH S.T., LAWSON J. (1980) Reference: The linguistic essential. Science, 210: 922–925.

SCHUMPETER, J.A. (1934) The theory of economic development: An inquiry into profits, capital, credit, interest, and the business cycle. Cambridge MA: Harvard University Press.

SMITH, A.D.M. (2005) The inferential transmission of meaning. Adaptive Behavior 13(4):311–324.

SZOSTAK, J., BARTEL D., LUISI P. (2001) Synthesizing life. Nature 409: 383–390

TAMARIZ, M. (2006) Evolutionary dynamics in language form and language meaning. Proceedings of the 6th International Conference on the Evolution of Language, 341–347.

TAMARIZ, M. VOGT P (in preparation) Synergistic complexification of linguistic forms and meanings through co-evolutionary dynamics.

TOMASELLO, M. (2003) Constructing a language: A usage-based theory of language acquisition. Cambridge MA: Harvard University Press.

WILLIAMS, G.C. (1992) Natural selection: Domains, levels, and challenges. New York: Oxford University Press.

WOESE, C.R. (1998) The universal ancestor. Proceedings of the National Academy of Sciences USA 95: 6854–6859.

WOESE, C.R. (2000) Interpreting the universal philo-genetic tree. Proceedings of the National Academy of Sciences USA 97(15): 8392–8396.

WOESE, C.R. (2002) On the evolution of cells. Proceedings of the National Academy of Sciences USA 99(13): 8742–8747.

www.ingramcontent.com/pod-product-compliance
Lightning Source LLC
Chambersburg PA
CBHW061006030426
42334CB00033B/3386